Postural Assessment

Hands-On Guides for Therapists

Postural Assessment

Hands-On Guides for Therapists

Jane Johnson, MCSP, MSc

The London Massage Company

Human Kinetics

Library of Congress Cataloging-in-Publication Data

Johnson, Jane, 1965-
  Postural assessment / Jane Johnson.
      p. ; cm. -- (Hands-on guides for therapists)
  Includes bibliographical references.
  ISBN-13: 978-1-4504-0096-1 (soft cover)
  ISBN-10: 1-4504-0096-5 (soft cover)
  1. Posture--Physiological aspects. 2. Mind and body. I. Title. II. Series: Hands-on
guides for therapists.
  [DNLM: 1. Posture--physiology. 2. Musculoskeletal Diseases--diagnosis. 3.
Musculoskeletal Diseases--prevention & control. 4. Musculoskeletal Manipulations. 5.
Postural Balance. WE 103]
  RA781.5J64 2012
  613.7'8--dc23

                                    2011013626

ISBN-10: 1-4504-0096-5 (print)
ISBN-13: 978-1-4504-0096-1 (print)

Copyright © 2012 by Jane Johnson

Acquisitions Editors: Loarn D. Robertson, PhD, and Karalynn Thomson; **Developmental Editor:** Amanda S. Ewing; **Assistant Editors:** Kali Cox and Brendan Shea; **Copyeditor:** Patsy Fortney; **Permissions Manager:** Dalene Reeder; **Graphic Designer:** Nancy Rasmus; **Graphic Artist:** Dawn Sills; **Cover Designer:** Keith Blomberg; **Photograph (cover):** Courtesy of Emma Kelly Photography; **Photographs (interior):** Courtesy of Emma Kelly Photography; **Photo Asset Manager:** Laura Fitch; **Visual Production Assistant:** Joyce Brumfield; **Photo Production Manager:** Jason Allen; **Art Manager:** Kelly Hendren; **Associate Art Manager:** Alan L. Wilborn; **Illustrations:** © Human Kinetics; **Printer:** United Graphics

Printed in the United States of America   10 9 8 7 6 5 4 3 2 1

The paper in this book is certified under a sustainable forestry program.

Human Kinetics
Website: www.HumanKinetics.com

United States: Human Kinetics
P.O. Box 5076
Champaign, IL 61825-5076
800-747-4457
e-mail: humank@hkusa.com

Canada: Human Kinetics
475 Devonshire Road Unit 100
Windsor, ON N8Y 2L5
800-465-7301 (in Canada only)
e-mail: info@hkcanada.com

Europe: Human Kinetics
107 Bradford Road
Stanningley
Leeds LS28 6AT, United Kingdom
+44 (0) 113 255 5665
e-mail: hk@hkeurope.com

Australia: Human Kinetics
57A Price Avenue
Lower Mitcham, South Australia 5062
08 8372 0999
e-mail: info@hkaustralia.com

New Zealand: Human Kinetics
P.O. Box 80
Torrens Park, South Australia 5062
0800 222 062
e-mail: info@hknewzealand.com

E5287

I dedicate this book to two groups of people. First, to the many hundreds of bodyworkers who have attended my postural assessment workshops. Thank you for standing in your underwear so that we may all learn more about posture and the similarities and differences among our bodies. Thank you also for your questions, which have kept me on my toes and interested in this subject for many years. Second, I dedicate this book to all of you who are new to postural assessment and for whom I have two pieces of advice: You have to start somewhere, and you don't have to agree.

Contents

Series Preface

Massage may be one of the oldest therapies still used today. At present more therapists than ever before are practicing an ever-expanding range of massage techniques. Many of these techniques are taught through massage schools and within degree courses. Our need now is to provide the best clinical and educational resources that will enable massage therapists to learn the required techniques for delivering massage therapy to clients. Human Kinetics has developed the Hands-On Guides for Therapists series with this in mind.

The Hands-On Guides for Therapists series provides specific tools of assessment and treatment that fall well within the realm of massage therapists but may also be useful for other bodyworkers, such as osteopaths and fitness instructors. Each book in the series is a step-by-step guide to delivering the techniques to clients. Each book features a full-colour interior packed with photos illustrating every technique. Tips provide handy advice to help you adjust your technique, and the Client Talk boxes contain examples of how the techniques can be used with clients who have particular problems. Throughout each book are questions that enable you to test your knowledge and skill, which will be particularly helpful if you are attempting to pass a qualification exam. We've even provided the answers too!

You might be using a book from the Hands-On Guides for Therapists series to obtain the required skills to help you pass a course or to brush up on skills you learned in the past. You might be a course tutor looking for ways to make massage therapy come alive with your students. This series provides easy-to-follow steps that will make the transition from theory to practice seem effortless. The Hands-On Guides for Therapists series is an essential resource for all those who are serious about massage therapy.

Whether you are a student or practitioner of physical therapy, osteopathy, chiropractic, sports massage, sports therapy or almost any form of bodywork, including fitness instruction, yoga and Pilates, you will no doubt appreciate the value in carrying out postural assessments as part of your overall assessment procedure. The assessment of posture helps you to determine whether there is any muscular or fascial imbalance and whether this imbalance might be contributing to a client's pain or dysfunction.

Being able to assess posture for imbalance is a skill required by most therapists, yet until now, there have been few resources to support practitioners in this task. *Postural Assessment* is intended as a comprehensive, user-friendly guide to this topic, with tips that will enable you to perform your observations in a confident and competent manner. This book focuses on what posture reveals about the relationships among various body parts so that you may be better informed about whether such relationships are causing or contributing to pain or discomfort or whether these relationships might affect the stability of joints by increasing or decreasing joint range. The emphasis is on the assessment of static posture in standing and seated positions. Written for complete beginners in the use of postural assessment as an examination tool, this book will generate ideas about what to look for, how to identify common postural forms and how to make sense of your observations.

Postural Assessment is organised into two parts. Part I explains how to get started and answers questions such as these: Why perform postural assessment? Who should have postural assessment? Where and when should postural assessment take place? What are the benefits of doing postural assessment? In this introduction you will learn about factors affecting posture and the ideal posture as well as how to provide the correct environment for postural assessment, what equipment you will need, how long it should take and how and why to document your findings. Line drawings illustrate the bony landmarks you will be using as you work through later chapters.

Part II contains the nuts and bolts of postural assessment—where you start and what you look for. Arranged according to whether you are observing your client from the back, side, front or in a seated position, each of the chapters in this part of the book follows the same format: A part of the body is presented with a line drawing and a short description of how to assess that body part, followed by a paragraph explaining what your findings mean. For example, what does it mean that your client has inwardly rotated shoulders—which muscles are short and tight, and which are long and weak? The What Your Findings Mean sections explain what various postures suggest about underlying musculature; they will help you relate your observations to real-life clients.

Notice that throughout chapters 3, 4, 5 and 6 (the chapters that make up part II), I refer to what each posture suggests, might mean, indicates, may mean or could mean. This is because I have chosen to analyze posture based on traditional assumptions regarding how muscles work. As a practitioner myself, I have inevitably weighted some suggestions more than others, based on my own experiences. It seemed helpful to include these sections to provide a starting point from which to develop your own ideas. In my experience, bodyworkers are markedly varied in their approaches to both the assessment and treatment of clients, and so you may not agree with my comments in these sections. Many steps offer quick tests you can do to justify these statements. I openly encourage you to question what your findings mean and ask how you could confirm for yourself, for example, that a person standing with protracted scapulae may have lengthened and weakened rhomboids and shorter anterior chest muscles. Perhaps more important, nowhere do I state what you should or could do to correct posture. This book is about what you observe, not what you do with your findings.

Putting together this text on postural assessment has been an interesting process. I was trained by educators who upheld a traditional view of musculoskeletal anatomy and physiology, and I have therefore held that knowledge at the core of my practice. However, it pays to keep an open mind and to be alert to progress in the fields of anatomy and physiology, perhaps even to question traditional assumptions about the way individual muscles work. I encourage you to seek out information to support your own conclusions regarding postural assessment. For example, in their article 'Anatomy and Actions of the Trapezius Muscle', Johnson, Bogduk, Nowitzke, and House (1994) carried out a study that 'refutes misconceptions about the actions of trapezius and how they are conventionally depicted' (p. 44). In another more recent article in a series on the psoas muscle, Thomas Myers (2001) questions whether the psoas is a hip flexor and whether this muscle contributes to thoracolumbar rotation.

The appendix contains a full set of charts for documenting your findings. These charts provide space for you to write your observations as you work through each step of postural assessment, whether you are observing your client from the front, side, back or in a standing or seated position.

As with other titles in the Hands-On Guides for Therapists series, the step-by-step format is supported throughout with tips to get you up to speed with the assessment technique. There are also Quick Questions at the end of each chapter so you can test yourself as you proceed. Unlike other titles in the series, *Postural Assessment* uses predominantly simple line drawings rather than photographs to facilitate identifying anatomical relationships.

As therapists, we are encouraged to work holistically with our clients, linking one body part with another and appreciating how body systems interrelate. Yet when attempting to process a new skill, we can sometimes feel overwhelmed trying to assimilate all the information. I am therefore hoping you will forgive me for compartmentalising the body and its assessment in the presentation of this book. Here, assessment procedures are presented in bite-sized chunks to give you time to process the information. Whatever your profession, if you are involved with helping people to look after and feel better in their bodies, I hope you will find this book of value.

Acknowledgements

Initial thanks must go to John Dickinson, former acquisitions editor at Human Kinetics, for believing my idea for a book on postural assessment had merit. Thanks also to Loarn Robertson for also liking the idea and accepting the formal proposal.

I am indebted to the many subjects who volunteered to be photographed for this book, including those whose images did not appear in the final version. As you may realise, one has to select material from a large number of photographs, and without these to choose from, this book would not have been possible. To everyone who hung around in their underwear that long evening in London, I thank you for your generosity.

I would like to give special thanks to Bruce and Patricia Robertson, my parents, whose photographs demonstrate the effects of postural change with increasing years. Also, to Siva Rajah, who agreed to having her photograph taken at very short notice, and to Tatina Semprini, for helping so much on the day of the photography shoot. Thanks to the photographer, Emma Kelly, who executed the photography shoot with diligence and humor, following my specifications to the letter. I hope your job as a police officer will not prevent us from working together again.

I extend my gratitude also to the following people who read and provided comments on the manuscript: Cameron Reid BSc(Hons), DO, osteopath, and fourth-year osteopathy students Zara Valentine and Jason Bianchi.

I would like to thank two people who helped in the creation of the cover photograph: Richard Lewis for agreeing to be the model and Sifu Andrew Sofos of the SAS Martial Arts Academy, who generously allowed us to film in his beautiful studio space here in London.

Last, I would like to acknowledge the Human Kinetics staff who helped put the book together: Loarn Robertson, Amanda Ewing, Brendan Shea, Kali Cox, Nancy Rasmus and Dawn Sills.

Getting Started With Postural Assessment

Do you have questions concerning the assessment of posture? The two chapters in part I set the scene for postural assessment. Chapter 1 addresses questions such as these: What is posture? What factors affect it? We examine the controversial question of whether there is an ideal posture before exploring who should have a postural assessment, when, where and why. The information in chapter 2 will help you to prepare for your first assessment. You will learn how long the assessment might take, what equipment you might use, key points regarding standard, or ideal, postural alignments, plus tips on documenting your findings. Both chapters conclude with Quick Questions, which you can use to test your understanding of the topics covered.

Introduction
to Postural Assessment

Welcome to *Postural Assessment.* Let's start by addressing some common questions concerning this fascinating subject. For example, what do we mean by the term *posture,* and why should you do a postural assessment? For which clients might postural assessment be helpful? When and where should the assessment be done? Do you need any equipment? This chapter answers these questions and examines factors affecting posture. You might wish to glance at the five Quick Questions at the end of this chapter before you begin, because doing so may help you retain some of the information if you are new to this subject.

What Is Posture?

Ask anyone to demonstrate poor posture, and it's a fair bet that most will adopt a slouched or hunched position, protracting their shoulders and rounding their backs to exaggerate the kyphotic curve in the thoracic spine. Ask for a demonstration of good posture, and most people automatically straighten up, raise their chins, and retract and depress their shoulders in a military-type attitude. Clearly, for most people, the term *posture* describes an overall body position, the way we hold ourselves or position our bodies, intentionally or unintentionally. Used in an artistic context, it might describe a pose, or a position held deliberately for aesthetic effect.

Good posture requires a person to maintain the alignment of certain body parts; poor posture is often acknowledged as a cause of musculoskeletal pain, joint restriction or general discomfort. When used in the context of therapy—physiotherapy, massage therapy, osteopathy or chiropractic, for example—the term *posture* more precisely describes the relationships among various parts of the body, their anatomical arrangement and how well they do or do not fit together. Bodyworkers have become familiar with postural terms such as *scoliosis* and *genu valgum,* which are used to describe a congenital,

Assessing Posture

Here are three people. One has had chest surgery, one has hypermobile joints and one has very tight external hip rotators. Can you tell which is which by looking at these photographs?

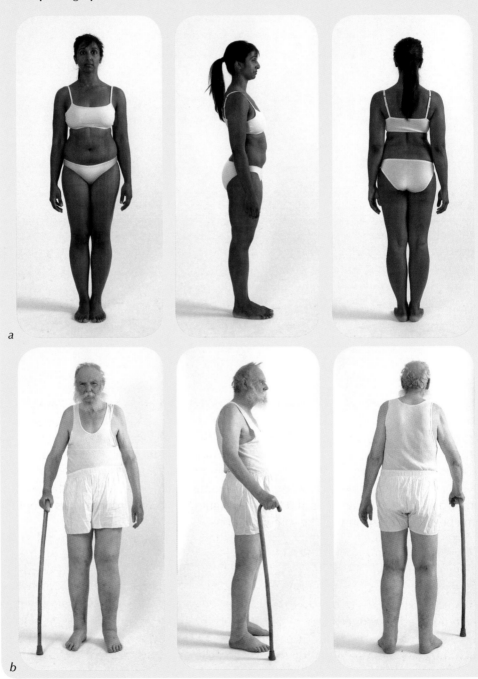

a

b

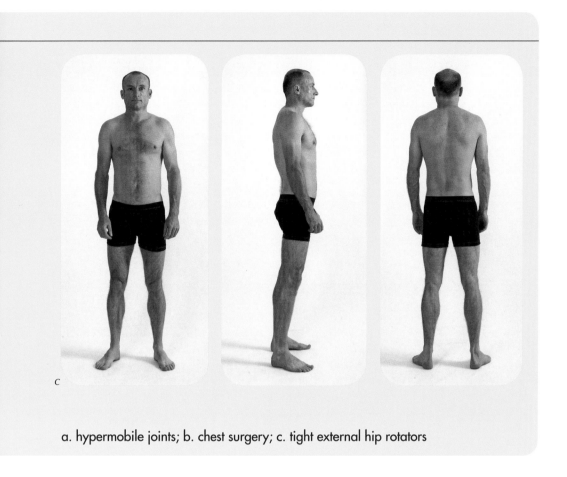

c

a. hypermobile joints; b. chest surgery; c. tight external hip rotators

inherited position, plus used to describe a position assumed through habit, such as increased thoracic kyphosis resulting from prolonged sitting in a hunched position.

Of course, the postures we assume provide clues to not only the condition of our bodies—traumas and injuries old and new, and mild or more serious pathologies—but also how we feel about ourselves—our confidence (or lack of it), how much energy we have (or are lacking), how enthusiastic (or unenthusiastic) we feel, or whether we feel certain and relaxed (or anxious and tense). Intriguingly, we all almost always adopt the same postures in response to the same emotions.

Observe 10 people feeling confident, motivated, and optimistic, and you will notice that most are standing tall, with their chests out and heads up, and that most have adopted a wide stance, giving themselves a wide base of support. They may be smiling or have a countenance that reflects their positive feelings. By contrast, observe 10 people feeling anxious, demotivated and pessimistic, and you may notice that they have shifted their weight to one leg, reducing their base of support (making them less stable), and that they stoop or flex at the waist, looking to the floor rather than up and ahead. They may touch the chin with one hand the way we sometimes do when we are thinking, and may even cross one or both arms against the chest in a protective manner.

If you are a teacher, you can demonstrate emotional postures to your class. Select one negative emotion and one positive emotion. Ask your class to act as if they were feeling extremely worried (or fearful or anxious or angry). It is important that all class members act out the same emotion. Observe what they do and what postures they adopt for a minute or so. Next, ask them to act as if they had just received a piece of fantastically good news. Again, observe what happens. Be sure to select the positive emotion as the second scenario to avoid students' retaining any sense of negative emotion. Also, suggesting that students carry out this exercise with their eyes closed prevents them copying one another. It is striking to observe how the majority of people adopt the same postures in response to the same emotions.

Although this book focuses on helping you to analyse the physical aspects of clients' postures, it is worth remembering that the postures we adopt reveal more than just the simple alignment of body parts. Our supposedly non-tangible, emotional states are inherently linked to our tangible, physical forms.

What Factors Affect Posture?

It is useful to consider the factors that affect posture to identify which ones you might be able to modify as a therapist and which might be better addressed by the client. There may be some factors affecting posture about which neither you nor the client can do anything at all. Table 1.1 provides examples of factors that affect posture. Perhaps you can think of additional items?

Is There an Ideal Posture?

You need only spend a short time comparing cadaveric specimens to discover that, anatomically speaking, we are not all the same. We may all have two scapulae, but the coracoid process projects at different angles, and the acromion process is similarly varied. The spinous processes of vertebrae are not all angled to the same degree, and some of us have one (or more) limb bones longer than the other, or feet or hands considered disproportionately large to the rest of our bodies, not to mention the variation in soft tissues. It is therefore not surprising that the physiological compensations needed to keep us in the upright position vary. So in answer to the question, 'Is there an ideal posture?' we can say, yes, there is an ideal, but with the caveat that it is not ideal for us all.

Traditionally, students of physiotherapy, osteopathy, and chiropractic have learned to assess posture by comparing the posture of their patients against images of an upright skeleton (see figure 1.1). Observations are made posteriorly, laterally and anteriorly to see how well the bits of a patient fit together compared to how the bits of the androgynous skeletal images fit together. Not surprisingly, the postures of many of us are observed to vary from those represented by the skeletal images.

An excellent early reference for the assessment of posture was *Posture and Pain* (Kendall, Kendall, and Boynton 1952), in which the ideal posture was referred to as 'standard posture'. The authors were quick to point out that such a posture should form the basis for comparison and was not an average posture. For the purpose of

Table 1.1 Factors Affecting Posture

Factor	Examples
Structural or anatomical	■ Scoliosis in all or part of the spine ■ Discrepancy in the length of the long bones in the upper or lower limbs ■ Extra ribs ■ Extra vertebrae ■ Increased elastin in tissues (decreasing the rigidity of ligaments)
Age	Posture changes considerably as we grow into our adult forms, with postures in children being markedly different at different ages.
Physiological	■ Posture changes temporarily in a minor way when we feel alert and energised compared to when we feel subdued and tired. ■ Pain or discomfort may affect posture as we adopt positions to minimise discomfort. This may be temporary or could result in long-term postural change if the position is maintained. ■ Physiological changes that accompany pregnancy are temporary (e.g., low backache before or after childbirth), but sometimes result in more permanent, compensatory postural change.
Pathological	■ Illness and disease affect our postures especially when bones and joints are involved. Osteomalacia may show up as genu varum; arthritic changes are often revealed when joints in the limbs are observed. ■ Pain can lead to altered postures as we attempt to minimise discomfort (e.g., following a whiplash injury a client may hunch the shoulders protectively; abdominal pain may lead to spinal flexion). ■ Mal-alignment in the healing of fractures may sometimes be observed as a change in bone contour. ■ Certain conditions may lead to an increase or a decrease in muscle tone. For example, someone who has suffered a stroke may have increased tone in some limbs but decreased tone in others. ■ As elderly adults, we tend to lose height as a result of osteoporotic changes and so develop stooped postures; postmenopausal women may develop a dowager's hump.
Occupational	Consider the postural differences between a manual worker and an office worker, and between someone active and someone sedentary.
Recreational	Consider the postural differences between someone who plays regular racket sports and someone who is a committed cyclist.
Environmental	When people feel cold they adopt a different posture to that when they feel warm.
Social and cultural	People who grow up sitting cross-legged or squatting develop postures that are different from those of people who grow up sitting on chairs.
Emotional	■ Usually, the posture we subconsciously adopt to match certain moods is temporary, but in some cases it persists if the emotional state is habitual. Consider the posture of a person who is grieving, or the muscle tone of a person who is angry. ■ Clients who fear pain may adopt protective postures.

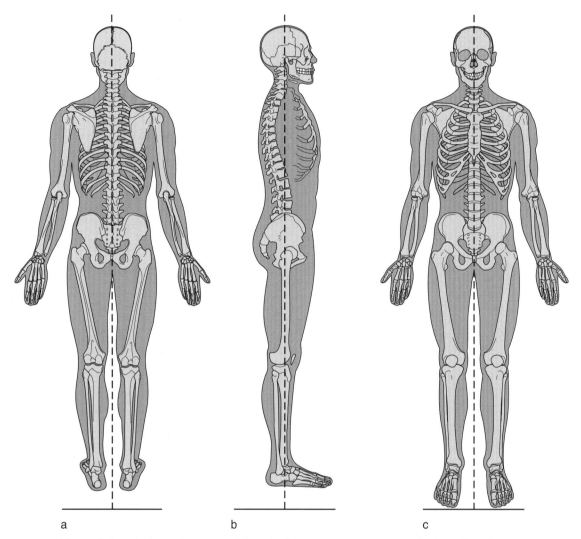

Figure 1.1   Traditional images of the ideal posture: *(a)* posterior, *(b)* lateral and *(c)* anterior.

understanding posture, using such images is as good a place to start as any. One of the disadvantages to this approach is that it may lead to a compartmentalisation of the body: a therapist might observe a client's neck to be excessively lordotic compared to the ideal posture, for example, and conclude that the problem is in the neck. Yet we need to take a broad view of clients and their bodies to identify the factors causing or contributing to their problems, because where a client experiences pain, discomfort or restriction in movement is not necessarily the source of the problem. By focusing too much on how local structures fit together, we risk overlooking what keeps them together. We tend to become focused on the part rather than the whole.

Let's say you have noted that a client has an excessively lordotic neck, with no history of disease or trauma. The posterior muscles may be shortened and tight, whereas the anterior muscles are lengthened and weak. So you surmise that stretching out the cervical extensors will improve the client's head position. Yet for many clients, the problem in the neck is likely to persist if postural imbalances in other areas of the body are not also addressed. An excessive lordosis in the cervical region of the spine often accompanies an increased kyphosis in the thoracic region. If the kyphosis remains unaddressed, the increased cervical lordosis is likely to persist despite treatments aimed at correcting it. This is because to compensate for a kyphotic spine, we must alter the position of the head (and ultimately the cervical spine); otherwise, we end up with our eyes facing the floor.

It is not just the positioning of body parts or the alignment of bones that we need to consider when examining posture. In recent years much has happened to improve our understanding of fascia and how it connects one part of the body to another. Advocates of techniques such as myofascial release argue that imbalances at one end of the body can affect structures at the other end because restrictions in fascia cannot be viewed as solely localised phenomena. For those interested in the importance of fascia to postural assessment and correction, I recommend *Fascial Release for Structural Balance* (Earls and Myers 2010) and *The Nature of Fascia* (Schleip 2008).

There is no doubt that an appreciation of the role of fascia in postural assessment will make you a better assessor. Yet because fascia embraces us entirely, linking all parts, there is no beginning and no end for the observer. I therefore hope that those of you already skilled at taking into account fascial tension in various parts of the body when you perform your postural assessments will forgive me for reverting to the more traditional compartmentalised approach. We do, after all, have to start somewhere.

Why Should I Do a Postural Assessment?

One of the first questions you might ask when considering postural assessment is why you would want to do it in the first place. What's the rationale behind such an assessment? The main reasons for carrying out a postural assessment are to acquire information, save time, establish a baseline, and treat holistically. Let's take each of these points in turn.

Acquire Information

First, and most important, performing a postural assessment gives you more information about your client. Here are three examples to illustrate this point:

- **Example 1.** Working with the general population, you have your fair share of clients suffering from back and neck pain. Many clients believe that their 'terrible posture' is due to the sedentary nature of their work, the long hours they spend slumped at a desk or driving. It would be helpful to know whether a client's pain does indeed stem from the adoption of habitual postures, or whether it might be due to something else. By distinguishing among various causes, you are more likely to be able to determine whether a change in working posture might be beneficial.

- **Example 2:** You are treating clients who regularly engage in sport or physical activity. A 30-year-old man comes to you complaining of recurring knee pain. He is a keen runner. Could his pain be aggravated by the posture of his lower limbs? Could he be flat-footed, have genu valgum or a leg length discrepancy, factors postulated to contribute to knee pain in runners? You observe your client, and his posture seems fine. Is it then more likely that his knee pain is the result of the quality or quantity of his training? Sometimes doing a postural assessment helps you rule *out* anatomical causes.

■   **Example 3:** Assessing a 49-year-old woman for worsening shoulder pain, you notice a decrease in shoulder muscle bulk during the postural assessment. One possible explanation for atrophy of the shoulder muscles (accompanied by a progressive decrease in range of movement) in a client with no history of trauma is adhesive capsulitis. The information you have gained from your observation has contributed to the formulation of your diagnosis, which may later be substantiated or refuted with the appropriate tests. It is important to remember that postural assessment is only one component of the assessment procedure, and that to make a diagnosis of any condition, all components of the assessment procedure need to be considered, along with current guidelines. For example, to support a diagnosis of adhesive capsulitis, you may follow guidelines such as those set out by Hanchard and colleagues (2011).

The postural assessment is also an opportunity to clarify observations about marks on the skin such as scars. Experienced clinicians know that clients sometimes forget to mention significant operations (such as appendectomies), being so used to the scar and having forgotten about the operation. Adults who received treatment for fractures in childhood may fail to mention this, either because they have forgotten about the incident or because they are not sure of its relevance to the problem they now have. Noticing old scars and mentioning them is a good way to get extra information that in some cases proves relevant.

Save Time

A second reason for carrying out a postural assessment is that in the long run it saves time. It may reveal facts that are pertinent to the client's problem that might otherwise have taken longer to establish. The relationships among body parts are more difficult to assess when someone is lying down to receive a treatment, but suddenly become obvious when they stand. Here are two examples:

■   **Example 1:** You are a sports massage therapist treating a typist who is normally fit and healthy. She is complaining of right-side anterior shoulder pain. Performing both the standing and sitting postural assessments, you observe that your client has a considerably protracted right scapula, something you had not noticed when your client was the prone position, a position in which both scapulae naturally protract.

■   **Example 2:** Your client is an elderly man with pain in his left ankle. Observing his posture from the posterior and anterior views, you get the impression that he does not bear weight equally through his lower limbs but seems to favour his left leg despite this being the problem ankle. There is slightly more bulk in the left calf muscle, too. Upon questioning, the client recalls fracturing his right ankle as a child and admits to feeling fearful about bearing weight through this ankle. Even though the client knows the fracture is fully healed, he reports always having relied more on his left leg for support. Could this information explain the pain in the client's *left* ankle? Could he have an arthritic ankle, or could the pain simply be due to the accruement of stresses in the joint from increased weight bearing? The subtle increase in muscle bulk you observe on the left calf in standing is something that you may not have spotted when the client was in the prone position or when performing range of movement tests. Observing the alteration in weight bearing has provided you with a significant piece of information.

Establish a Baseline

A third reason for performing a postural assessment is that it helps you to establish a baseline—a marker by which you might judge the effectiveness of your treatment. If your client has muscular pain in the low back resulting from the position of the pelvis, and you prescribe exercises and stretches to correct this posture, you will no doubt need to reassess the client at some stage to determine whether there has been any change in the pain and whether this can be attributed to an alteration in the position of the pelvis. Many therapists use subjective feedback from the client as a benchmark measure of effectiveness. If we suspect that a problem *is* the result of poor posture, we need to identify whether we have made any impact (directly with massage and movement, or indirectly with prescribed exercises and stretches) on the client's upper body posture. The way to do this is to assess posture before and after the treatment intervention. For more information on this subject, see the section When Should Postural Assessment Be Done? on page 13.

Treat Holistically

Finally, it could be argued that by including an analysis of posture as part of our assessment, we are offering a more complete service, in keeping with the idea of treating people holistically, not compartmentalising them as a bad knee, a frozen shoulder, or whiplash. We keep records of clients' states of health and physical activities, so it seems logical that we also keep a record of their postures.

Now that you are aware of the many good reasons for carrying out a postural assessment, we need to identify those clients for whom a postural assessment would be most beneficial.

Who Should Have a Postural Assessment?

This book is aimed at those of you working with the general population—whether fit and active or unfit and sedentary—rather than with clients who may be hospital based and suffering from sudden trauma or long-term illness. Ideally, you should perform a postural assessment on all clients presenting for sports or remedial massage, physiotherapy or osteopathy treatments. If you are working as a fitness professional—one of your aims being to strengthen weak muscles—or as a teacher of yoga—aiming perhaps to lengthen muscles—you too will find postural assessment beneficial because it will help you identify muscle imbalances and you can therefore design the most effective exercises and postures for your clients. However, with some clients, a postural assessment may not be appropriate, such as the following:

- An anxious client
- A client unable to stand because of pain or illness
- A client who is unstable when standing or when getting to or from the standing position
- A client who does not understand the purpose of the assessment or who does not give consent to having one performed
- A client with a condition that would benefit from a different form of assessment

When working with an anxious client, you may want to postpone a postural assessment while you develop a rapport. Once that is established, you can carry out a more thorough assessment, including that of posture. It would be inappropriate to assess the posture of a client who is unable to stand because of pain or illness. Remember, you can still assess a client in a seated position (see chapter 6, page 125). In some cases a postural assessment is warranted but must be performed with care. For example, you may want to assess an elderly person who has suddenly become unbalanced when using a regular walking aid. In this case you need to assess the patient standing with the aid, yet you must also ensure safety. Similar caution needs to be taken when assessing a client with a recent injury. With such patients—particularly those with injury in the lumbar spine, pelvis or lower limbs—weight bearing or a change in posture may aggravate discomfort. Some clients may be unsettled by how close you are to them during a postural assessment; with such clients, you should clearly explain your intention and the purpose behind the assessment.

Postural assessment can be very useful in hospital settings. For example, it would be helpful to assess a patient who has suffered the sudden trauma of a stroke. However, because such a patient is likely to have an abnormal increase or decrease in muscle tone, the section What Your Findings Mean within this text will not apply. Those sections address the effects of postural imbalance on healthy people rather than those suffering from sudden trauma. Similarly, using this book to assess the posture of a person suffering a degenerative condition such as Parkinson's disease will provide useful information, but not as useful as information gained from using tools specifically designed for people suffering from this condition. Clients suffering conditions that affect the normal functioning of the nervous and muscular systems are better assessed using assessment tools specifically designed for the assessment of their conditions.

Where Can Postural Assessment Take Place?

Because postural assessment usually requires that clients be in their underwear, a warm clinic or treatment room is the best location. However, important information can be gained about a client who sits for long periods of time by observing her seated posture at work. In this case, the person will be clothed. Although this will not provide as much information as when you can see the position of the client's joints, it gives a useful overall impression of whether musculoskeletal pain may be due to poor seated posture.

When Should Postural Assessment Be Done?

Postural assessment is usually performed following the consultation and once you have important information about your client's medical history. It is important to take the medical history first because something may come to light that affects whether you perform the postural assessment. For example, if a client reports having dizzy spells when he stands too long, you might carry out the assessment more quickly than usual, or have an assistant on hand and a chair close by. Or you might decide not to do the assessment at all.

Some therapists like to reassess posture following each treatment, but of course, this depends on how often you are seeing that client and the nature of your treatment. Your intervention may be a one-off treatment, in which case you need to do the assessment before and after treatment, or it may comprise a series of treatments and home care exercises over a period of time. In this case, it would be more appropriate to assess posture initially and then later, once you and the patient have had a chance to implement the treatment plan. It may not be necessary to assess the entire posture, or even to do a postural assessment at all. However, if you are keen to see whether intensive treatment and exercise have relieved the problem you believe to be the result of poor posture, you will certainly want to reassess your client at some stage.

Closing Remarks

This chapter has introduced you to the concept of postural assessment and some of the factors affecting it. Hopefully, it has also answered some of your initial questions about this topic. Chapter 2 prepares you to undertake your first assessment.

Quick Questions

1. What are three factors that affect posture?
2. What are two reasons for doing a postural assessment?
3. What are two examples of when a general postural assessment (as described in this text) might not be appropriate?
4. In most cases, why is it important to take a medical history before carrying out a postural assessment?
5. When analysing various parts of the body and how they fit together, why is it important to always take an overall view of the client?

Preparing for Postural Assessment

Let's now prepare to carry out the assessment. In this chapter you will find information on how to carry out the assessment, the equipment required, how long it should take, plus an outline of what to look for in general terms. You will also find a more detailed description of the standard (ideal) postural alignments, images of which you should hold in your mind when carrying out your assessments as you work through subsequent chapters. Here, too, are ideas for how to document your findings and some general cautions and safety issues to consider before you start.

Equipment Required

The following equipment is useful when learning to carry out postural assessments:

- A warm, private room
- A full-length mirror
- Body crayons (and cleansing wipes)
- Postural assessment charts
- A model skeleton

TIP It is a good idea to have your own posture assessed, perhaps by a colleague working through the steps in this book. In this way, you can experience what it feels like to be the subject of the assessment and will be better informed when performing the procedure yourself.

Body crayons are helpful to have at hand. Crayons used for children's face painting are inexpensive and readily available from most stores selling party supplies. They are also usually non-toxic and hypoallergenic, but do check with your client before using them. You might want to start by practising on family and friends. Use the crayons to mark bony landmarks on the client to help you judge the distances of these structures from the midline of the body, the angles of bones and joints, and the relationships they form with one another. Try marking the following points, which are all on the back of the body:

- Medial border of the scapula
- Inferior angle of the scapula
- Spinous processes of the spine
- Olecranon process of the elbow
- Posterior superior iliac spine (PSIS)
- Knee creases
- Midline of the calf
- Midline of the Achilles tendon

If you choose to use crayons, remember to have some cleansing wipes at hand to remove your marks.

It is helpful to have some kind of chart on which to record your observations and to serve as a prompt so you know what to look for. Sample charts are provided for posterior, lateral and anterior assessments in the appendix (page 141). These match the step-by-step assessments in chapters 3, 4, 5 and 6. Finally, having a full-size model skeleton at hand will be helpful for reminding you of anatomical structures, their size and their placement.

 If you are using a model skeleton that is supported on a base stand (rather than hanging), it is likely to be constructed with a central pole running through the bodies of the vertebrae. This means that the spine of your model does not depict the normal lordotic and kyphotic curves we all have; in your model the curves are flattened.

Time Required

If you are new to postural assessment, you may find that you need at least 30 minutes to do a full assessment including anterior, posterior and lateral views. With practice, you can assess the whole body, from all angles, in five minutes or less. A skilled practitioner carries out a general postural assessment quickly, noting only very obvious deviations from the standard posture. When a client presents an ongoing or unresolved problem, or when you are assessing as part of rehabilitation, you will need to take greater care to establish whether postural imbalances are contributing to the problem or are likely to become problematic in the future. In this case the assessment may take longer.

Postural Assessment Steps

Once you have established that a postural assessment will be useful and that the client understands what this involves, you are ready to begin. The client should stand in her underwear, and have hair tied up if it obscures the face and neck. It is a good idea for female clients to wear a normal bra rather than a sport bra because sport bras can obscure the scapulae and spine at the back, making observation of this part of the body more difficult. Because you want to observe the everyday stance of your client, instruct her to adopt her usual standing posture rather than have her position her feet in any particular way.

To begin, hold in mind some general questions regarding your client's body. Table 2.1 provides some suggestions. The questions here are designed as prompts and are not exhaustive.

Most people feel vulnerable when asked to stand unclothed to have their bodies analysed in this way. Yet this is an assessment procedure that usually takes place at the start of the therapeutic relationship. How such a procedure is conducted can be critical

Table 2.1   General Questions Regarding Your Client's Body

Overall stance	▪ Is the weight equally distributed through the lower limbs or shifted to one side? ▪ Does the client look balanced or unbalanced? ▪ Does the client appear to be swaying forwards, backwards or to one side?
Alignment of body parts	▪ Does each component of the body seem balanced with relation to other parts? ▪ Is the head centred over the thorax? ▪ Is the thorax centred over the pelvis? ▪ Are the limbs equidistant from the trunk?
Bones	▪ Do bones appear normally shaped? ▪ Do any bones appear misshapen, bowed or damaged?
Joints	▪ Do joints appear to be in their neutral, resting positions, or is there any mal-alignment? ▪ Do any joints appear swollen?
Muscles	▪ Is there equal bulk on the left and right sides of the body? ▪ Is there noticeable hypertrophy or atrophy anywhere? ▪ Does there appear to be an increase or a decrease in muscle tone anywhere?
Skin	▪ Are there any areas of inflammation, discolouration or dryness? ▪ Are there any scars, blemishes or bruising?
Physical attitude	▪ Does the client look comfortable? ▪ Does the client appear to be able to maintain the posture with ease? ▪ Are there any areas of tension?

in determining whether trust and rapport are established between the patient and the therapist. Although you have your own bedside manner, and will approach this form of assessment in a way that matches your individuality as a practitioner, it does not hurt to remember the importance of making your observations as non-judgemental as possible.

Observations that to you may seem commonplace and a matter of fact may have huge emotional significance for the patient. Acknowledge these emotions, and take care to make your observations sensitively. Working supportively and without criticism is as important during the postural assessment as they are during the treatment, rehabilitation and education components of your work. Best results are achieved when clients feel safe and calm, perhaps even curious about your observations of their posture. Without their trust and confidence, you will never get a really true picture of their posture.

Standard Alignments

As you learned in chapter 1, in this book you are going to compare your observations with a traditional image of good postural alignment. However, rather than describe this as the ideal alignment, we're going to adopt the term used by Kendall, Kendall and Boynton (1952) and refer to it as *standard* alignment. The illustrations in this book have traditionally been used as a benchmark against which to make observations regarding posture. These images have been chosen as representative of optimal body alignment because in this upright position compressive forces are distributed evenly over the surfaces of joints.

The alignment of joints as shown in these images does not require any increase in tension in soft tissues, and little work is required by muscles other than to correct postural sway. Deviations from this optimal joint alignment increase the stress on ligaments and require that muscles exert more effort not just in the associated joint but usually in the joints above and below the affected part, as part of the body's corrective mechanism. If mal-alignment persists— that is, deviations from the positions illustrated here—detrimental structural changes may occur.

With joints aligned in this 'ideal' way, ligaments counter any tendency for joints to flex or extend. Where there are no ligaments to do this, muscles fire intermittently to counter any joint deviations. The muscles increase their energy expenditure only if the joint falls very anterior or very posterior to the plumb line, or line of gravity, or if they are required to maintain a position of imbalance indefinitely. An excellent and comprehensive text describing the anatomy and function of all joints is *Joint Structure and Function* (Levangie and Norkin 2001). For information on the static and dynamic control of posture, see chapter 13 of that text.

The figures in the sidebar shown on pages 19, 20 and 21 outline where a plumb line (shown as the vertical black line in the illustrations) should fall with respect to various parts of the body, and provide general observations for when postures are said to be good, or ideal. Illustrations within this sidebar show standard alignments for the posterior, lateral and anterior views of the body. If you wish to skip this section, you could turn to the detailed step-by-step instructions explaining what to observe in each of these views in chapters 3, 4 and 5.

Standard Alignments

Standard Posterior Alignment

Head

Plumb Line
Through midline of the skull

General Observations
The head should be facing forwards with no rotation and no lateral flexion

Shoulders

Plumb Line
Equidistant between the medial borders of the scapulae

General Observations
The height of the shoulders should be approximately level. However, the shoulder of the dominant hand may be lower than the shoulder of the non-dominant hand.

Pelvis and thigh

Plumb Line
Through the midline of the pelvis

General Observations
•The posterior superior iliac spines (PSIS) should be equidistant from the spine and be level.
•The greater trochanters of the femurs should be level.
•The buttock creases should be level and equal.

Knees and legs

Plumb Line
Between the knees

General Observations
•The legs should be straight and equidistant from the plumb line with no genu varum or genu valgum.
•Calf bulk should be equal on the left and right legs.

Neck

Plumb Line
Through midline of all cervical vertebrae

General Observations
The neck should appear straight with no lateral flexion.

Upper limbs

General Observations
•The arms should hang equidistant from the trunk, palms facing the sides of the body.
•The elbows should be level.
•The wrists should be level.

Thorax and scapulae

Plumb Line
Through midline of all thoracic vertebrae

General Observations
•The scapulae should be equidistant from the spine, the medial borders of each approximately 1.5 to 2 inches (3.8 to 5 cm) from the spine. The scapulae should lie flat against the rib cage with no anterior tilting. The inferior angles of the scapulae should be level, with no evidence of elevation, depression, or scapular rotation.
•Flare in the rib cage should be symmetrical left and right.

Lumbar spine

Plumb Line
Through midline of all lumbar vertebrae

General Observations
The lumbar spine should be straight with no curvature to the right or left.

Ankle and feet

Plumb Line
Between the medial malleoli

General Observations
•The lateral malleoli should be level.
•The medial malleoli should be level.
•The Achilles tendon should be vertical.
•The calcaneus should be vertical.
•The feet should be turned out slightly.

Standard Lateral Posture

Head

Plumb Line
Through the earlobe

General Observations
The head should appear positioned over the thorax—neither pushed forwards with chin out nor pulled back.

Shoulders

Plumb Line
Through the shoulder joint: specifically, through the acromion process (not shown on this illustration)

General Observations
The shoulders should be neither internally nor (in rare cases) externally rotated.

Lumbar spine

Plumb Line
Through the bodies of the lumbar vertebrae

General Observations
The lumbar spine should have a normal lordotic curve that is neither exaggerated nor flattened.

Knees and legs

Plumb Line
Slightly anterior to the knee joint

General Observations
There should be neither flexion nor hyperextension at this joint in standing.

Neck

Plumb Line
Through the bodies of most of the cervical vertebrae

General Observations
• The cervical spine should have a normal lordotic curve that is neither exaggerated nor flattened.
• There should be no deformity at the cervicothoracic junction such as a dowager's hump.

Thorax and scapulae

Plumb Line
Midway through the trunk

General Observations
• There should be a normal kyphotic curve in this region that is neither exaggerated nor flattened.
• The chest should be held comfortably upright and not excessively elevated (military posture) nor depressed.

Pelvis and thigh

Plumb Line
Through the greater trochanter of the femur

General Observations
• The pelvis should be in a neutral position. That means the anterior superior iliac spine (ASIS) is in the same vertical plane as the pubis.
• The ASIS and the PSIS should be approximately in the same plane. There should be no anterior or posterior pelvic tilt.
• Gluteal and thigh muscle bulk should appear equal on both the left and right sides.

Ankle and feet

Plumb Line
Slightly anterior to the lateral malleolus

General Observations
There should be normal dorsi-flexion at the ankle.

It is important to remember that although the plumb line in the lateral view should run vertically through the earlobe and bodies of most cervical vertebrae, when being set up for use as a marker, it is positioned slightly anterior to the lateral malleolus and *not* against the earlobe, cervical vertebrae, acromion or other structures listed here. Remember, this is an ideal posture, showing where, ideally, the plumb line ought to bisect the body in such a way that equal portions of the body appear anterior and posterior of the plumb line.

Standard Anterior Alignment

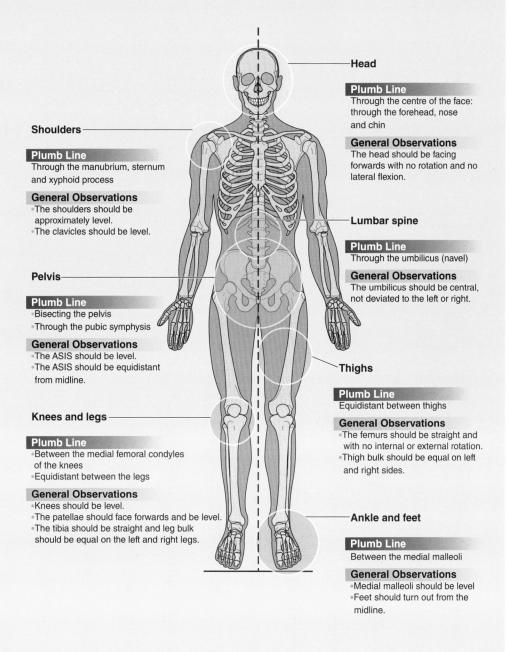

Head

Plumb Line
Through the centre of the face: through the forehead, nose and chin

General Observations
The head should be facing forwards with no rotation and no lateral flexion.

Shoulders

Plumb Line
Through the manubrium, sternum and xyphoid process

General Observations
• The shoulders should be approximately level.
• The clavicles should be level.

Lumbar spine

Plumb Line
Through the umbilicus (navel)

General Observations
The umbilicus should be central, not deviated to the left or right.

Pelvis

Plumb Line
• Bisecting the pelvis
• Through the pubic symphysis

General Observations
• The ASIS should be level.
• The ASIS should be equidistant from midline.

Thighs

Plumb Line
Equidistant between thighs

General Observations
• The femurs should be straight and with no internal or external rotation.
• Thigh bulk should be equal on left and right sides.

Knees and legs

Plumb Line
• Between the medial femoral condyles of the knees
• Equidistant between the legs

General Observations
• Knees should be level.
• The patellae should face forwards and be level.
• The tibia should be straight and leg bulk should be equal on the left and right legs.

Ankle and feet

Plumb Line
Between the medial malleoli

General Observations
• Medial malleoli should be level
• Feet should turn out from the midline.

The Plumb Line

The plumb line represents the line of gravity; it is a vertical line drawn from the body's centre of gravity to within the body's base of support. The human centre of gravity is just anterior to the second sacral vertebra. Although the plumb line is included in these illustrations, many who carry out postural assessments believe that practitioners ought to focus more on the relationship between one body part and another, rather than on the relationship between a particular body part and the plumb line.

Traditionally, a plumb line has been used in the analysis of posture by positioning the client against it in such a way that the plumb line falls equidistant from the medial malleoli when viewing a client anteriorly or posteriorly, and anterior to (about 1 inch, or 2.5 cm, in front of) the lateral malleolus when viewing the client laterally. To ensure that their clients were positioned correctly, practitioners placed the feet of their clients on a card or plinth that had the outline of two feet painted on it. However, this approach has been criticised because it could result in clients being forced to adopt foot positions that are abnormal for them. In reality, a client may have a naturally wider (or narrower) stance than the painted foot positions allow, or may stand with his feet turned out or in to a greater or lesser degree than the foot positions allow. Although the use of a plumb line (without forced foot positions) is valuable in carrying out research into posture, and specialist practitioners may use it (along with gridded background paper to measure the alignment of body parts on the horizontal plane), in practice, it is not necessary and may even be cumbersome.

TIP   Traditionally, the client's head is supposed to sit nicely above the thorax with the plumb line falling through the ears and the acromion. However, because the shoulder joint is a mobile structure, if the scapula is protracted (as is often the case), the head may appear to be in the wrong position relative to the plumb line when, in fact, it is the scapula that is out of alignment.

Identifying the 'standard' seated posture is difficult because it does not exist. Seating needs to accommodate the needs of the client and what she is doing while seated. Images such as the one shown in figure 2.1 recommend that the hips be flexed to a certain degree, but this assumes that the client is sitting at a desk and not, for example, driving a train, using the toilet or eating a meal. Generally speaking, it is accepted that posture is less likely to be compromised when the head is positioned over the thorax, the lumbar spine is supported with the knees comfortably flexed, and the feet are flat on the floor.

Documenting Your Findings

As you perform your postural assessment, you are going to want to document your findings. There are many ways to do this. The most common method is simply to write out your findings in longhand. By complete contrast, you could speak your findings

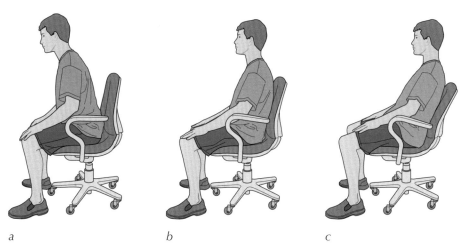

a b c

Figure 2.1 Some common seating recommendations. Positions *a* and *c* should be avoided, and position *b* places less stress on the spine and soft tissues.

Neutral Pelvic Positions

Before moving on, I want to clarify what I mean by a neutral pelvis. A neutral pelvis is one in which the left and right iliac crests, the left and right posterior superior iliac spines (PSIS) and the left and right ischia are level when viewed posteriorly, as in figure 2.2.

When viewed from the side, the ASIS and the pubis should be approximately in the same plane (Anderson 1978), as in figure 2.3.

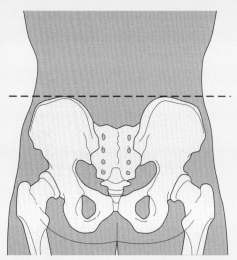

Figure 2.2 Neutral pelvis viewed posteriorly.

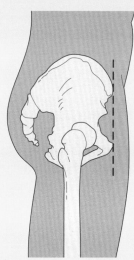

Figure 2.3 Neutral pelvis viewed laterally.

into a recorder, or if carrying postural assessment as part of research, you could ask your participant to sign a release form granting you permission to take photographs that you later analyse. Audio-recording information is quick and easy, but it is likely to be distracting for the client to whom you may not wish to reveal your findings just yet. Photographs enable you to take your time over the assessment, but they are no substitute for having the client in front of you. Many participants were photographed for selection of images used in this book. One of the advantages of assessing posture in this way was that I was able to take my time with each photograph to identify those deviations in posture that would best support the information in the text. One of the disadvantages is that a photograph retains only a certain amount of clarity and is not life size.

How you record information depends on which method is most appropriate for you. A visually impaired therapist, for example, may 'observe' a client by palpation. The therapist may assess the relationships among various body parts using her hands while audio-recording her findings to transcribe later. In some clinical settings therapists are required to document their findings using standardised forms. Clinical trials often require clinicians to record deviations from the standard postural views using measuring devices such as rulers and goniometers. In everyday practice, most therapists simply note whether any deviations they observe are mild, moderate or marked. Often, arrows or shading are used to indicate deviation of a body part or an increase in muscle tone. To get started, you will find postural assessment charts in the appendix.

Cautions and Safety Issues

There are few contraindications to carrying out a postural assessment. It is not advisable if a client has pain on prolonged standing (or prolonged sitting if you are doing the seated postural assessment in chapter 6). Clients with low blood pressure may get dizzy if asked to stand for too long, so have a chair at hand. Take care when examining clients who can stand without discomfort but who feel unbalanced. This is particularly common among elderly people or clients who are recovering from an injury or surgery to the lower limbs and have only recently started fully bearing their weight.

Make sure your clients are appropriately disrobed before you start the assessment. The elderly man on page 4 preferred to keep his vest top on because he had recently suffered a fall and was badly bruised. If he were being assessed for the bruising, I would have asked him to remove his top, but I knew that asking him to do this himself while standing and with a walking stick in one hand was dangerous.

Finally, if you are using face crayons to mark up your client's body, you need to check that your client is not allergic to them. Allergic reactions to these crayons are rare. However, if you see increased blood flow to the skin where you are using the crayon, remove the marks you've made.

Closing Remarks

Now that you are fully prepared, it's time to carry out your first assessment. Choose either the posterior, lateral or anterior view, and turn to the appropriate chapter for your step-by-step guide.

Quick Questions

1. What are four useful pieces of equipment to have when carrying out a postural assessment?
2. What bony landmarks are useful to identify before starting a posterior postural assessment?
3. What are five general questions to ask yourself concerning the client's overall stance, alignment, bones, joints, muscles, skin and physical attitude?
4. What is a neutral pelvis?
5. What are some possible contraindications to postural assessment?

Carrying Out Postural Assessments

In part II you will find step-by-step information that will enable you to carry out postural assessments posteriorly (chapter 3), laterally (chapter 4), anteriorly (chapter 5) and with your client in a seated position (chapter 6). Starting with the head and neck and working down through the shoulders, thorax, arms, lumbar spine, pelvis, thighs, legs and feet, each chapter tells you what to look for and explains what variations in your findings might mean. Packed with thought-provoking questions, these chapters will help you identify which muscles are likely to be shortened and tight and which are likely to be lengthened and weak. Use the illustrated postural assessment charts in the appendix to help you document your findings while carrying out the assessments. Keep in mind what you learned about making general observations in chapter 2 as you work through the more detailed material here.

As noted in the preface, the What Your Findings Mean sections contain information based on commonly held beliefs about how muscles function as well as my own experiences. You will no doubt notice that these sections include questions and phrases such as *may mean, could indicate* and *might suggest* rather than sweeping statements of fact. You should also note that many postures may have more than one cause, or result from a combination of factors. For this reason, postural observation should form only part of your assessment procedure. You will likely also be carrying out muscle length tests and range of movement tests, and palpating your subjects to confirm diagnoses.

A number of good textbooks contain more information on these subjects. For example, excellent texts for information on all aspects of joint testing are *Orthopaedic Physical Assessment* (Magee 2002) and *Management of Common Musculoskeletal Disorders* (Hertling and Kessler 1996). *Muscle Testing and Function* (Kendall, McCreary, and Provance 1993) not only describes how to assess muscles, but also contains much information about posture. An excellent handbook for quick reference regarding joints is *The Clinical Measurement of Joint Motion* (Green and Heckman 1993). Also, *Joint*

Structure and Function (Levangie and Norkin 2001) provides superb descriptions and illustrations of normal joint positioning and factors affecting this.

The What Your Findings Mean sections are included because students especially need some form of guidance for the systematic analysis of posture. You may be of the opinion that the position in which a joint rests, for example, is entirely due to anatomical factors, or you may believe that how we hold our bodies is influenced by how we feel, the emotional states we perpetuate. I am not asking you to agree with my suggestions regarding these findings, but rather, to use them as a starting point on which to build the case for your own analysis.

To illustrate the postures in this book, multiple photographs were taken of 18 subjects. Each subject was asked to stand naturally and then filmed from the anterior, lateral and posterior views. None were told how to stand, and footprints were not used to indicate where they should place their feet. Where part of a photograph has been used to illustrate a particular step within the text, it is worth remembering that this may not be the *only* aspect of that subject's posture that is noteworthy. As you work through the following chapters, examine and compare the photographs to see whether you can identify characteristics in addition to those described in the text.

Posterior Postural Assessment

Let us get started with your first assessment. In this chapter you will learn to assess your client's posture posteriorly. Although it is wise to observe a client's overall posture, guiding your eyes from the head to the toes to get a general feel for symmetry and balance, when you are first learning how to carry out postural assessments, it is helpful to compartmentalise the body, observing each segment in turn. Working through each of the steps here and answering the associated questions will teach you how to perform a thorough posterior postural assessment. Each step includes a section called What Your Findings Mean with tips on spotting shortening or lengthening of muscles. After reading this chapter, you will have an insight into what may be causing the imbalances you observe. These imbalances may be contributing to a pain, discomfort or joint restriction.

First, locate the posterior postural assessment chart in the appendix on page 142. This chart corresponds to the steps you are going to read through in this chapter: 17 for the upper body and 14 for the lower body, making a total of 31 steps. You may carry out the steps in any order, although it is logical to follow the order in which they are presented here, from head to toe. It is best to use the chart once you have finished reading this chapter and are ready to perform your first postural assessment.

Having read chapter 2, Preparing for Postural Assessment, you are ready to assess your client. He should be standing comfortably in a warm room, perhaps facing a mirror and with his back to you. Experienced practitioners are able to carry out a thorough postural assessment (posteriorly, laterally and anteriorly) in 5 to 10 minutes. However, it may well take you longer than this when you are first learning, so practise on family and friends. Work through each of the steps listed here, and take a break if your client needs to sit down or starts to get cold.

TIP Postural assessments are carried out with clients in their underwear. Although it is important to be able to observe the body as a whole, when first learning postural assessment, it is useful to assess the top half of the body first, and then the lower half. This way, your clients can retain some of their clothes and may feel slightly more at ease, especially if they have not had a postural assessment before.

For this section, female clients are assessed wearing a bra, and male clients with a bare torso. It may be difficult to complete all of the steps if a female client is wearing a sport bra with a T-bar at the back because these often obscure the thoracic spine and inferior angle of the scapulae.

TIP Chatting with your clients as you carry out their postural assessments can put them at ease. However, keep in mind that when they are replying to you, they may try to turn or tilt their heads, which will alter your findings, especially when carrying out steps 1 through 5 concerning the head, neck and shoulders.

STEP 1   Ear Level

The first thing to look at is the level of your client's ears. Are the earlobes level? If your client has short hair, you will easily be able to see the ears; clients with long hair will need to tie their hair up and out of the way. If you cannot see the client's neck, simply leave this section of your assessment blank.

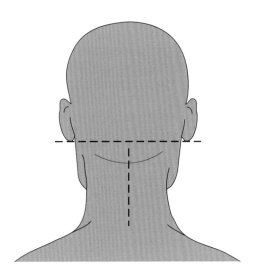

TIP   Some clients instinctively offer to hold their hair up and out of the way. Avoid having them do this because it alters the position of the head, neck and shoulders, which you need to observe in a neutral position (i.e., standing relaxed with their arms by their sides).

What Your Findings Mean   Uneven ear height could suggest that your client has her head tilted to one side, with a laterally flexed cervical spine. Lateral flexion of the neck can result from shortened muscles on the side to which the neck is flexed. For example, if the head is tilted to the right, the upper fibers of the trapezius may be tight on that side, as might the right levator scapulae, right sternocleidomastoid, and right scalene muscles. Less commonly, some clients have one ear positioned slightly higher on one the side of the head than on the other, and do not necessarily have a laterally flexed neck. Sometimes such clients are aware of this and report finding it difficult to get glasses or sunglasses to fit properly.

STEP 2   Head and Neck Tilt

This is similar to step 1 and may be used instead of step 1 if you cannot see the client's ears. Here you are observing whether the head is tilted to one side and asking, Is there any lateral flexion in the neck?

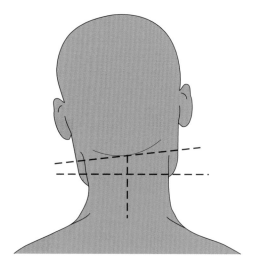

What Your Findings Mean   As in step 1, if the head appears tilted to one side, this indicates tightness in the muscles that laterally flex the head and neck, on the side to which the neck is tilted. For example, if your client were laterally flexed to the left, the left levator scapulae, sternocleidomastoid, and scalene muscles would all be tight, as would the upper fibers of the trapezius on the left. Clients with shoulder pain often flex their necks to the side of the pain, subconsciously, to minimise movement and reduce discomfort.

TIP   If your client is suffering from torticollis, there will be marked lateral flexion. Also known as 'wry neck', this is a spasming of the neck muscles resulting in lateral flexion, rotation of the neck or both. It is common following whiplash.

STEP 3   Cervical Rotation

Next, check whether there is any rotation in the cervical spine. Is your client looking straight ahead, or is the head rotated slightly to the right or to the left?

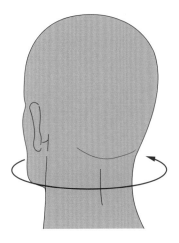

TIP   One way to assess for rotation of the head is to ask yourself whether you can see more of one side of the client's face than the other. Can you see more of the eyelashes on one side? Or more of the cheek perhaps?

These two people both believe they are standing with their heads facing forwards. If you look closely, you can just slightly see more of each person's jaw on one side. The man is rotated to the right (you can see more of the right side of his jaw), whereas the woman is rotated to the left (you can see more of the left side of her jaw). This is subtle but could be relevant if the client were reporting neck pain.

What Your Findings Mean   As you know, many muscles contribute to rotation of the head, including sternocleidomastoid and scalenes. The woman shown above could therefore have a tight right sternocleidomastoid muscle, tight left scalenes, plus a tight left levator scapulae muscle relative to these same muscles on the opposite side of the neck.

STEP 4   Cervical Spine Alignment

Is your client's cervical spine straight? This step is similar to steps 2 and 3, but it concerns the spine rather than the position of the head and neck. Look at the extensor muscles of the neck: Is there any increased tone on one side? Technically, a postural assessment involves only observation. However, in practice, it is useful to include some elements of palpation when you are first learning. (See also steps 2 and 3 of the lower body postural assessment on pages 51 and 54, which describe the need to palpate the iliac crest and posterior superior iliac spines of the pelvis.) A simple way to assess the alignment of the cervical spine with your client standing is to gently palpate the spinous processes and mark these using a body pen or crayon. Then stand back and observe your marks.

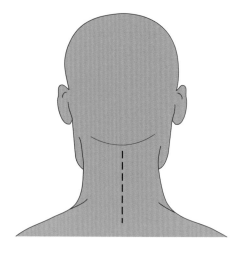

TIP When palpating the cervical spine, it helps to place one hand gently on the forehead of your client simply to stabilise the head and neck.

Palpation of the cervical spine can be tricky for several reasons. First, the spinous processes of some of the cervical vertebrae are bifurcated and approximate each other when the neck is in a neutral position as in standing, making them difficult to palpate individually. Second, they all lie deep to the ligamentum nuchae. Third, the extensor muscles of the neck are active when the client is standing so you have to palpate through thick tissue and active muscles.

TIP If you are using body pens, place your marks where you feel the spinous processes to be when you palpate, not where you think they ought to be. Sometimes with palpation, a vertebra feels slightly too far to the left or too far to the right of the midline, and it is tempting to mark where you think it ought to be, in the midline of the neck. However, what you have discovered could well be a misaligned vertebra and should be recorded.

What Your Findings Mean   Few of us have perfectly straight spines. Standing back to observe the marks you have placed on your client, you may see that all but one of the vertebrae appear to be in alignment. Knowing that a vertebra may be out of alignment is useful for massage therapists in particular, because this factor could be contributing to the client's problem and is a good example of when referral to an experienced physical therapist, osteopath or chiropractor may be appropriate.

TIP If the cervical spine does not appear to be straight, be careful how you reveal this information (if at all) to the client. Many clients could become worried if told they have mal-aligned cervical vertebrae. Remember, many of us go about our daily lives with less than straight spines and have no pain and no problems in this structure whatsoever. The purpose of putting marks on the neck area is simply to help you identify deviations when learning the technique of postural assessment. Such deviations may provide additional information that could inform your treatment.

STEP 5   Shoulder Height

Now let's look at your client's shoulders. Are they level, or does one appear higher than the other?

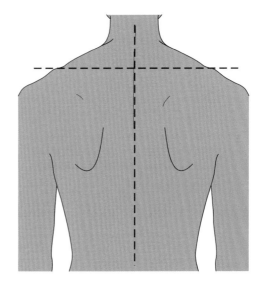

What Your Findings Mean   Shortening in levator scapulae and the upper fibers of the trapezius may contribute to one shoulder appearing higher than the other. If a scapula is elevated, you would expect the inferior angle of that scapula to be superior to the inferior angle of the scapula on the opposite side. Here is an interesting question: How do you know whether one shoulder is truly higher or the other is lower? Try this simple exercise: shrug your shoulders, elevating your scapulae; then relax. Now depress your shoulders; then relax. Which movement did you find easier, elevation or depression? Most people find that shrugging the shoulders is easier than depressing them. It seems reasonable to assume that if your client's right shoulder appears higher, muscles on the right are shorter and tighter than the corresponding muscles on the left. An exception to this might be if you were assessing someone with a neurological condition (e.g., having suffered a stroke) and she had a dropped shoulder as a result of low tone on one side of her body.

Therapists have observed that, for many people, the dominant shoulder is naturally depressed and slightly protracted. If you are right-handed, your right shoulder may be slightly lower and more protracted than your left.

Clients with neck pain may subconsciously elevate their shoulder protectively in an attempt to reduce their discomfort. This woman is standing 'relaxed'. Observe how she holds her right arm. She has suffered neck pain in the past, but at the time this photograph was taken, and for many months previous to that, she was pain free. Would you agree that her right shoulder is elevated? Can you see also how her neck is also laterally flexed and slightly rotated to the right?

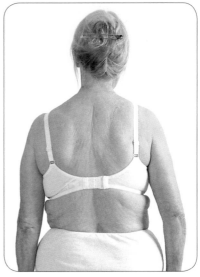

STEP 6   Muscle Bulk and Tone

Another thing you can look at is whether there is an increase or a decrease in muscle bulk on any part of the shoulders. You can document any increase or decrease in tone by hatching or shading the relevant illustration on your postural assessment chart.

What Your Findings Mean   Manual workers may have hypertrophy in muscles on the side they use to carry, lift or support heavy objects. Similarly, sportspeople may have an increase in muscle bulk on the dominant side. For example, right-handed archers often have hypertrophied rhomboids on the right because they use the right arm to draw back the bow, contracting the rhomboids maximally to retract the scapula. Conversely, observation of people with adhesive capsulitis or who have had their upper limb immobilised may reveal atrophy of the shoulder muscles on the affected side.

TIP   Disuse of the shoulder results in atrophy in all of the associated muscles, something that is particularly apparent in older clients who are often low in body fat and have less muscle bulk than younger clients. You can often tell whether a client is not using one shoulder (perhaps following injury) by observing the supraspinatus and infraspinatus muscles because these appear noticeably atrophied on the side of the injury.

STEP 7   Scapular Adduction and Abduction

Next, take a look at the scapulae and their relationship to the client's spine. Observing the relationship between the medial borders of the scapulae and the spine, decide whether the scapulae are adducted (retracted) or abducted (protracted). Many clients, unless engaged in regular exercise or sporting activity involving the upper body, have slightly protracted scapulae. This could be due in part to the kyphotic posture many people adopt when sitting.

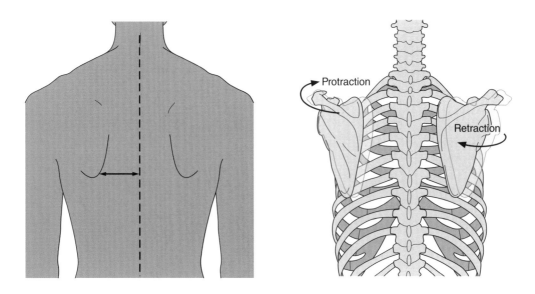

TIP   If you cannot see the medial border, gently palpate for it. To locate it, ask your client to place his hand behind his back while you do this. Remember that, in doing so, the scapula will change position. You may find that drawing a horizontal line on the skin directly down this border helps you get a better idea of the position of the scapulae.

What Your Findings Mean   Protraction of the scapulae often accompanies poor posture in which the rhomboids and the lower fibers of the trapezius are lengthened and weak bilaterally. Retraction of the scapulae is much less common and occurs when people adopt a military-style posture: chests pushed up and out, shoulders drawn back and down. In this case rhomboids might be shortened on both the left and right sides of the body. Clients engaged in sporting activity in which retraction predominates on one or both sides of the body (e.g., javelin throwers and archers) might demonstrate unilateral shortness in the rhomboids on the side of the retraction. Observation of clients who regularly engage in sporting activities involving bilateral retraction of the scapulae—such as rock climbing and rowing—may reveal hypertrophy in both left and right rhomboids.

Consider, also, what happens to the medial border when the scapulae rotates. With upward rotation the medial border and inferior angle are abducted from the spine, lengthening the rhomboid major and shortening the rhomboid minor and levator scapulae. With downward rotation, the medial border and inferior angle are adducted towards the spine, shortening the rhomboid major and lengthening the rhomboid minor and levator scapulae. Table 3.1 summarises this information. Notice that the serratus anterior has been included in this table because it attaches to the medial border of the scapulae on the anterior surface of the bone. For more information about rotation of the scapula, see step 9.

When assessing the shoulder region, as with any area of the body, be careful not to jump to conclusions regarding the source of shoulder pain. Just because a person stands with protracted scapulae and an internally rotated humerus, for example, does not mean that her scapular pain results from the anatomical positions of these bony structures. There are other possible sources of pain. For example, way back in 1959, Cloward reported on the likelihood of scapular and upper limb pain originating from cervical discs.

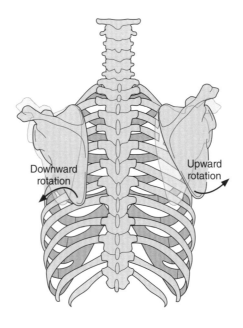

Table 3.1 Changes in Muscle Length Depending on Scapular Position

	Upward rotation	Downward rotation
Position of medial border	Both the medial border and the inferior angle are abducted from the spine.	Both the medial border and the inferior angle are adducted towards the spine.
Lengthened muscles	Lower fibers of the trapezius Rhomboid major Serratus anterior	Upper fibers of the trapezius Rhomboid minor Serratus anterior
Shortened muscles	Upper fibers of the trapezius Rhomboid minor Serratus anterior	Lower fibers of the trapezius Rhomboid major Serratus anterior

STEP 8   Inferior Angle of the Scapula

Still focusing on the scapulae, locate the inferior angle on each bone and compare their positions. Are they level with each other, or is one superior? Mark them with a body pen if you need to. Remember that scapulae can elevate and depress, as shown here. Look at the photo on page 35. Can you see that this person has an elevated right scapula?

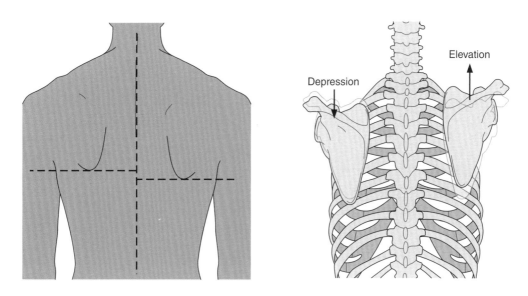

TIP   Again, if you cannot see the inferior angle and you find it difficult to palpate, ask your client to place her arm behind her back, which will make this part of the bone more prominent. Once you have located this part of the bone, be sure to ask the client to relax, returning her arms to her sides so you can observe this bony landmark.

What Your Findings Mean   The inferior angle is elevated when the whole scapula is elevated. Muscles of scapular elevation may be shorter on the side of the elevation. So if you observe that the inferior angle on the client's left scapula is higher compared to the inferior angle of the client's right scapula, this could mean that the upper fibers of the trapezius on the left, plus the left levator scapulae are shortened.

Before moving on to the next step, consider for a moment what you might observe on the anterior of the body in a client with an elevated scapula. When the right shoulder is elevated, you might expect to see the right clavicle raised too, because the two bones are attached at the acromioclavicular joint, as you know.

TIP   To demonstrate this relationship, hold your right clavicle with the fingers of your left hand, shrug your shoulders and feel what happens. Alternatively, observe your clavicles in the mirror as you shrug.

STEP 9   Rotation of the Scapula

Upward rotation describes upward movement of the glenoid fossa; downward rotation describes downward movement of the glenoid fossa. Using this figure to help you, are you able to determine whether your client has a rotated scapula?

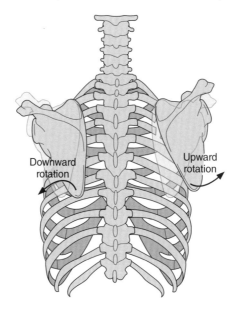

As you can see, if a scapula rotates, not only will the medial border move with relation to the spine, but the position of the inferior angle will change also. Therefore, differences in the position of the scapular border can be due to rotation of the scapula (not just to adduction or abduction of the scapula); differences in the position of the inferior angle can be due to rotation of the scapula (not just to elevation or depression of the scapula).

In reality, these movements do not occur in isolation. An upwardly rotated scapula will be adducted at the superior angle and abducted at the inferior angle; a downwardly rotated scapula will be abducted at the superior angle and adducted at the inferior angle. In both of these cases, the scapulae could be elevated or depressed. You can see why it is a good idea when starting out with postural assessment to break up your assessment into its component parts.

What Your Findings Mean   Scapular rotation occurs as a result of tension in some tissues and weakness or slackening in others. Upward rotation suggests tension in the levator scapulae, rhomboid minor and the upper fibers of the trapezius, and weakness in the rhomboid major and the lower fibers of the trapezius. Downward rotation suggests tension in the lower fibers of the trapezius and rhomboids major and weakness in the middle and upper fibers of the tapezius, the rhomboid minor and the levator scapulae. Refer to table 3.1 on page 38 for a summary of this information.

STEP 10   Winging of the Scapula

A term that gets bandied about quite a lot in relation to scapulae is *winging*. In addition to the six movements already described, scapulae can tilt against the rib cage in such a way that the inferior angle becomes prominent. The figure here illustrates tilting. This image shows a right lateral view of the spine with the scapula in the normal and tilted positions. Where the medial border of the scapula appears particularly prominent along with the inferior angle, this is sometimes referred to as winging. However, the term *winging* is more accurately used to describe what happens to the scapula when the serratus anterior muscle is unable to keep the scapula fixed against the rib cage so that it seems to stick out like a wing. This is not commonly observed.

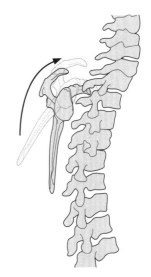

What Your Findings Mean   True winging can occur because the long thoracic nerve is damaged or the muscle itself is damaged. Obviously, this is not a condition you are likely to come across. However, if muscles attaching to the anterior of the scapula shorten, they could tilt the scapula forwards so that the inferior angle becomes more prominent. Your posterior postural assessment can provide clues as to the state of soft tissues on the anterior of the body.

Review of Scapular Movements

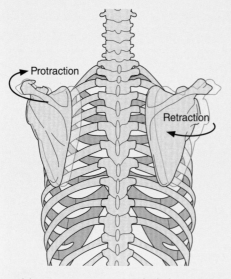

Adduction (retraction) and abduction (protraction).

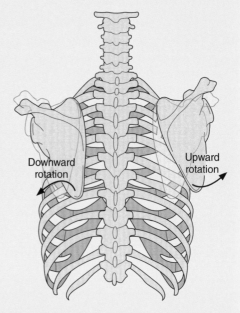

Upward and downward rotation.

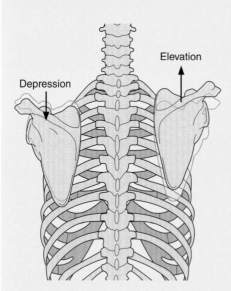

Elevation and depression.

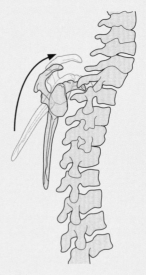

Tilt.

STEP 11   Thoracic Spine

Turn your attention now to the thoracic spine. Is it straight, or is there evidence of scoliosis? (The figure on the right shows evidence of scoliosis.) If necessary, palpate for spinous processes, and mark them as you did for the cervical spine, remembering that many of us have spines that deviate from the vertical position.

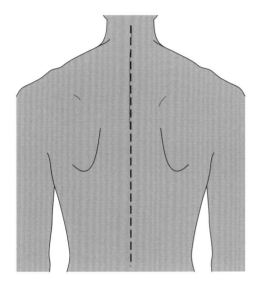 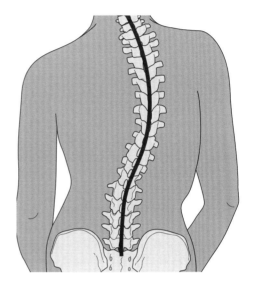

TIP   A trick you can do if you do not wish to use body pens is to gently run a fingernail down either side of the spine, just enough to leave a slight red mark but obviously not deep enough to scratch the client. Then, stand back and observe the track marks you have made: Are they straight, or do they deviate?

You will gain more information about the spine when you carry out your lateral postural assessment, in which you assess for lordosis and kyphosis. Nevertheless, it is useful when making your first impressions of this region to note whether the client is kyphotic or has a flat back.

What Your Findings Mean   It is important to remember that there are many causes of scoliosis. It may be congenital, the result of injury or altered biomechanics or the result of a leg length discrepancy, in which case the pelvis tilts laterally and the spine is forced to compensate. The treatment for each of these causes is unique, and you should not jump to the conclusion that treatments aimed at lengthening the shortened tissues on the concave side of the curve will resolve the problem. As with the cervical spine, overall curvature and deviations in individual vertebrae can help explain pain in this region. Discovering that your client has a laterally deviated spine is useful, yet you may not be able to treat this condition, or it may not need treating at all.

TIP   If you observe scoliosis, be cautious about revealing this information to the client. Some clients might be alarmed to discover that their spines are not straight. Furthermore, the fact that they have less-than-straight spines may have nothing at all to do with the complaint about which they have come to you.

STEP 12   Thoracic Cage

Take a look at the positioning of the thoracic cage. How does it sit in relation to the client's head and hips? Does it appear rotated, or perhaps shifted to one side?

TIP   One way to understand the relationship among the head, thorax and pelvis is to imagine them as three-dimensional blocks or cylinders that can move with respect to one another. The thoracic cage could be a cylinder positioned between the head (also a cylinder) and the pelvis (a rectangular block). Not only can these structures rotate, but they can also shift to one side, sliding across one another like a child's wooden play blocks.

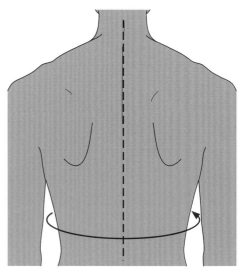

In this photo, the medial border of the right scapula appears not only more prominent than the left scapula but also closer to the observer. This suggests that the trunk is rotated clockwise.

What Your Findings Mean   Many muscles affect the rotation of the thorax, not just the muscles attaching to this part of the body. Table 3.2 summarises how muscle length may correspond to the positions of the trunk when it is rotated. If you are wondering why muscles of the neck are included in this table, look in a mirror and rotate your trunk one way; notice what the muscles of your neck must do in order to keep your head facing forward.

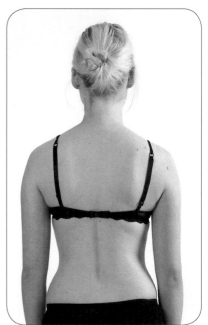

Table 3.2   Muscle Length Corresponding to Rotation of the Trunk

	Trunk rotated to the right	Trunk rotated to the left
Shortened muscles	Right internal oblique Left external oblique Left psoas* Left lumbar erector spinae Muscles that rotate the neck to the left	Left internal oblique Right external oblique Right psoas* Right lumbar erector spinae Muscles that rotate the neck to the right

*The psoas is not a definitive rotator, yet recent research suggests it may be more involved in stability of the spine, including rotation, than originally thought.

STEP 13   Skin Creases

Another useful observation you can make concerns whether there are more or deeper skin creases on one side of the trunk than the other. This step is not about whether there are or are not skin creases. After all, clients with low body fat may have no creases at all, whereas clients who are overweight may have many. This step is about determining whether there is a *difference* between the left and right sides of the body and enables you to use your observations of the skin to explain what may be happening to deeper structures.

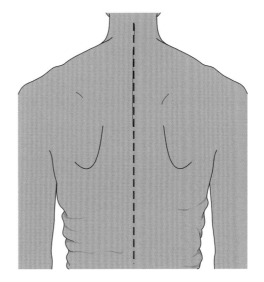

What Your Findings Mean   When we laterally flex, we lengthen tissues of the opposite side while compressing the side to which we are flexing. The result is to deepen the creases on the side to which we are flexed.

TIP   Ask your client to lean (laterally flex) to the right, and observe what happens to the skin creases on the right side of the trunk.

A key lateral flexor of the spine is the quadratus lumborum. More or deeper creases on the right side of the trunk may indicate a shortened quadratus lumborum on that side.

STEP 14   Upper Limb Position

Let's look now at the upper limb and compare the space formed between the client's arm and body. Is this the same on the left and right sides? Can you see how in both of these people the space between the left arm and the trunk is larger than that between the right arm and the trunk when they are standing in what they consider to be a relaxed posture?

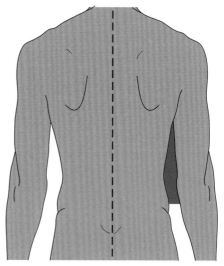

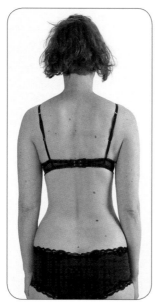
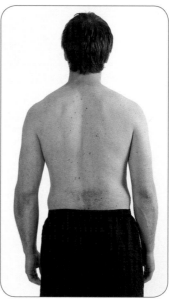

What Your Findings Mean   Here are three possible explanations for the postures demonstrated in these figures:

- The arm on the side showing greater space is abducted more. Could the supraspinatus or the deltoid (or both) be shorter than the corresponding muscles of the opposite shoulder?

- The client is laterally flexed to that side. If this is the case, the client may have a shorter quadratus lumborum on the side to which he is flexed.

- The client is hip hitched, and her pelvis is laterally tilted upwards on the side to which she is flexed.

TIP   Try this for yourself. Standing in front of a mirror with your arms resting at your sides, laterally flex to the right, letting your right arm hang loose. Notice that the space between your right arm and that side of your trunk increases.

STEP 15 Elbow Position

Take a look at your client's elbows. Observation of the elbow is useful for two reasons. First, whether or not the elbows are level (a) often ties in with whether the client has a dropped or elevated shoulder or is laterally flexed to one side. Second, observing the position of the elbow can help you assess whether the client is internally rotated (b) at the glenohumeral joint, as can the position of the client's hands (see step 16 on page 48). An internally rotated humerus might contribute to shoulder pain caused by the impingement of soft tissues.

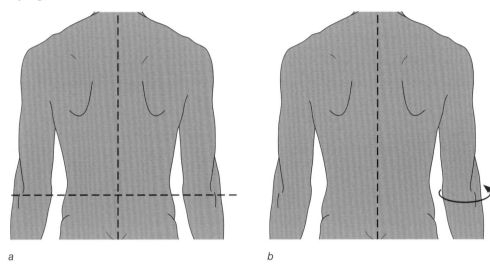

a   b

TIP Ask your client to flex his elbows and place a small dot on the olecranon process of each using a body crayon. Then, with the client standing with his arms relaxed, compare the left and right elbows. Ask the client to deliberately rotate one arm internally and observe how the elbow on that side moves laterally.

The left shoulder of this person appears internally rotated.

What Your Findings Mean The internal rotators of the humerus (such as the subscapularis, pectoralis major and teres major) might be shortened.

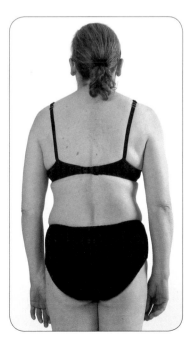

STEP 16   Hand Position

Observe the position of the client's hands and how much of the palms you can see.

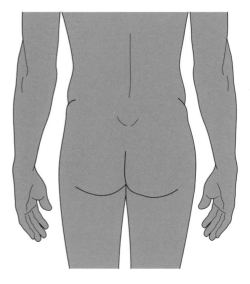

TIP   Remember that shortening of the supinators or pronators in the elbow or wrist, or both, could also alter the position of the hand in standing.

What Your Findings Mean   The more of the palm you can see, the more internally rotated the humerus is. Again, knowing that your client has an internally rotated humerus is useful because this can explain shoulder pain. Turn back to page 47 to see a good example of a person who demonstrates this posture.

STEP 17   Other Observations

Finally, before moving on to the lower body, make note of anything else you have observed that you have not yet documented, such as scars, blemishes or unusual marks on the client's skin. You might also note something obvious that affects posture, such as the fact that the client has an arm in a cast or sling, or an obvious swelling such as in the bursa of the olecranon. The man shown here had bad bruising on his back following a recent fall. He was wearing a vest top for the assessment, which initially obscured this observation.

To get the most from a lower limb postural assessment, have the client stand barefoot, his back to you, wearing underwear or running shorts.

STEP 1   Lumbar Spine

Is the lumbar spine straight, or is there evidence of scoliosis? Although an increase or a decrease in the lumbar curve is best observed laterally, note your first impressions of this area: does it appear lordotic or flattened?

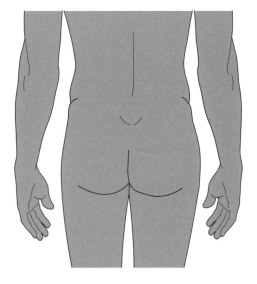

Look at the two photos here. Can you see that neither of these people has a straight spine? What do you observe about the skin creases on their waists? Can you see that the right crease is slightly deeper?

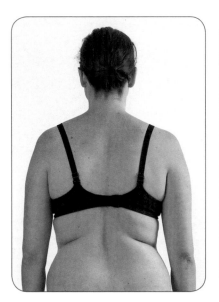
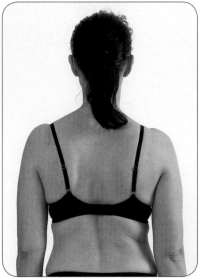

What Your Findings Mean   Curvature may indicate a variety of things including recent injury (such as a disc herniation), muscle spasm, scoliosis, muscle imbalance or lateral flexion due to the pelvis being raised on one side.

STEP 2   Pelvic Rim

Many experienced therapists believe that postural imbalances can be addressed by adjusting the position of the pelvis, that whether the imbalance is in the upper or lower body, repositioning of the pelvis into a more neutral position helps overcome these imbalances. The following steps (and those in the lateral assessment) relating to the positioning of the pelvis are considered important by some practitioners. Thus, check to see whether the pelvis is level or whether there is any lateral tilting.

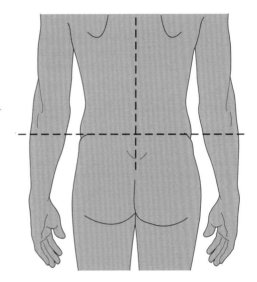

TIP   A good way to check whether the pelvis is level when you are new to postural assessment is to sit or crouch down behind your client and gently place your hands on her waist. Press first into the fleshy part of the waist and then down onto the bony iliac crest. Gauge whether the left and right sides of the pelvis feel level.

Illustration *a* shows a normal pelvis, and illustration *b* shows a pelvis laterally tilted upwards to the right and laterally tilted downwards to the left.

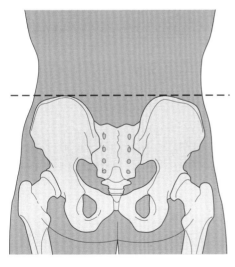

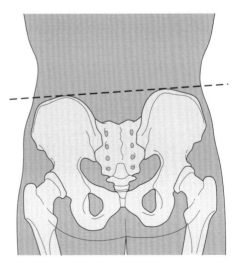

a   *b*

You can feel what a laterally tilted pelvis feels like by standing in front of a mirror, both feet on the floor. Imagine that you have your leg in a cast and cannot flex at the knee. Place your hands on your hips and slowly lift the heel of your right foot off the floor, but keep the toes of your right foot on the floor as you do this. You can see and feel the right side of your pelvis as it rises and as you laterally flex to the right at your lumbar spine to accommodate this position.

What Your Findings Mean To compensate for a pelvis that is raised on the right, a client may have increased lateral flexion of the lumbar spine (to the right), which may correspond with the appearance of more or deeper skin creases on the right. In this case, the right quadratus lumborum muscle may be shorter, as may some of the right lumbar erector spinae muscles. The hip joints are affected also. The right hip is adducted, whereas the left hip is abducted. Therefore, a client may have a pelvis raised on the right with shortened hip abductors on the left and shortened adductor muscles on the right.

TIP To help you visualise the effect a laterally tilted pelvis has on the hips, picture the pelvis as a tabletop with two table legs beneath it (Levangie and Norkin 2001). The legs are free to swing left and right (i.e., to abduct or adduct). Now imagine tilting the tabletop down to the left (up to the right). What happens to the table legs? They will continue to hang down perpendicularly, but notice what has happened to the angles they now form with the tabletop, (representing the attachment of the femur at the hip). The right leg is adducted (the internal angle has decreased), and the left leg is abducted (the external angle has decreased).

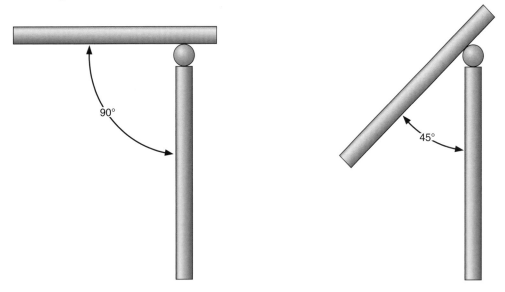

A client with one leg longer than the other may have a laterally tilted pelvis.

Now look at the ischium. Can you see that in our illustration it is elevated on the right? What might the consequences of this be for the length of the hamstring muscles? If the knee joints were level, could the left hamstrings be shorter than those on the right?

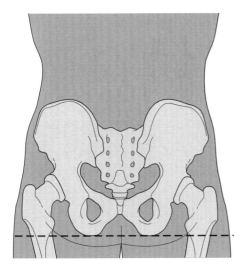 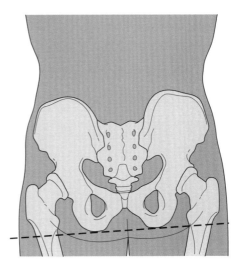

These findings are summarised in table 3.3.

Table 3.3   Possible Effects of a Laterally Tilted Pelvis

	Pelvis raised on the right	Pelvis raised on the left
Lumbar spine	Flexed to the right; concave on the right	Flexed to the left; concave on the left
Lumbar muscles	Shortened right quadratus lumborum and right lumbar erector spinae	Shortened left quadratus lumborum and left lumbar erector spinae
Effects on the hip joint	Right hip is adducted; left hip is abducted	Left hip is adducted; right hip is abducted
Effects on the muscles of the hip	Shortening of the right hip adductors and left hip abductors; imbalance between the left and right hamstrings	Shortening of the left hip adductors and the right hip abductors; imbalance between the left and right hamstrings

TIP The iliac crest *roughly* equates with the position of the fourth lumbar vertebra. This is useful information should you need to palpate this area of the spine.

STEP 3   PSIS

The posterior superior iliac spines are located directly beneath the dimples some clients have in this region. Placing your thumbs here and gauging whether the PSIS points are level are another way to confirm a lateral tilt of the pelvis in standing. In this photo, the position of the dimples suggests that the person's right PSIS is higher than his left PSIS. Do you think his spine is straight or laterally flexed to the right slightly?

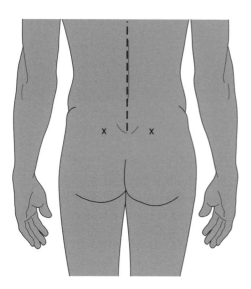

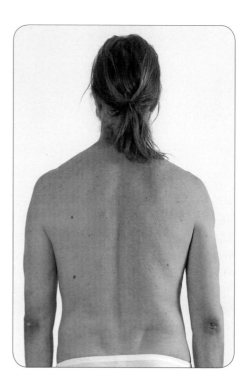

What Your Findings Mean   If you agree that the left and right PSIS should be positioned on the same horizontal plane, yet observe one to be higher, this suggests that the pelvis is laterally tilted.

STEP 4　Pelvic Rotation

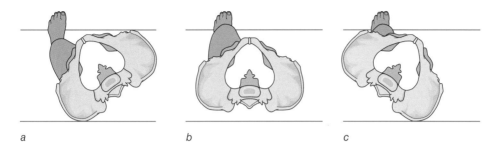

a　　　　　　　　　　　　b　　　　　　　　　　　　c

(a) Pelvis rotated anti-clockwise (to the left).

(b) Normal pelvis.

(c) Pelvis rotated clockwise (to the right).

Like a cube-shaped bead on a piece of string, the pelvis can rotate with respect to the spine in the way that the bead can rotate with respect to the string. With your hands on the pelvis of the client, see if you can determine whether the pelvis is rotated with respect to the lumbar spine. (You will need to check this by viewing your client both laterally and anteriorly using the charts in the appendix.) Is it rotated clockwise, in which case the right side of the pelvis will be closer to you and the left side away from you? Or is it rotated anti-clockwise, in which case the left side of the pelvis will be closer to you and the right side farther away. The illustrations above exaggerate rotator movements; in reality, they are far more subtle.

TIP　To determine pelvic rotation, it helps to imagine that the client is standing between two sheets of glass, one in front and one behind. Does the PSIS on one side of the pelvis appear to be closer to the glass behind the client than that on the other side?

What Your Findings Mean　If the pelvis is rotated away from you on the left (clockwise), the right internal oblique and left external oblique may be shortened. If the client has a pelvis rotated forwards to the right (anti-clockwise), the opposite may be true.

　　Pelvis rotation also affects the feet and knees. Please refer to chapter 5 for more on this.

STEP 5 Buttock Crease

It is not always possible or appropriate to observe the crease of the buttock, where it meets the proximal thigh. If the client is wearing long shorts or cycling shorts, you will not be able to see these creases.

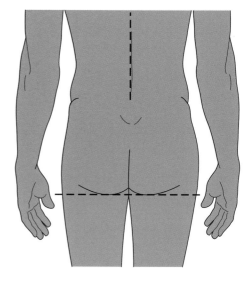

TIP The buttock crease is formed by the fat overlying the buttock muscle and does not indicate the location of the inferior fibers of the gluteus maximus muscle.

Some therapists choose to palpate the ischial tuberosities to check whether they are level as an indication of whether the pelvis itself is level. However, if you are not familiar with palpating this area, you may decide that doing so is too invasive or not inappropriate at this stage of your assessment.

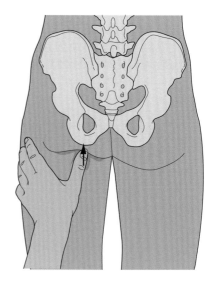

The woman in this photograph is a good example of a person with uneven buttock creases. Observe the position of her underwear, too. Does it look like the right side of her pelvis is higher than the left?

What Your Findings Mean   Clients who bear weight more on one side of the body than the other may have a deeper buttock crease on that side. This is also often true of clients with laterally tilted pelvises. So a client with a pelvis tilted upwards on the right, as in step 2, might appear to have a deeper left buttock crease. Could differences in the height of the buttock creases also correspond with leg length discrepancies? The following figures illustrate the appearance of varying bone lengths in the lower limb with respect to the buttocks.

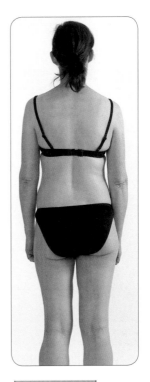

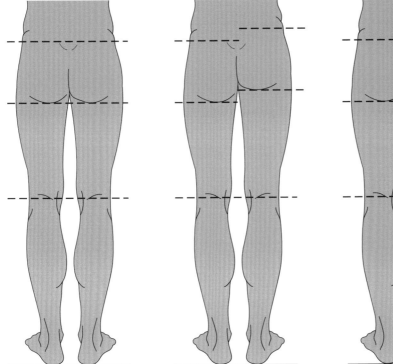

| Normal | Right femur longer | Right tibia longer |

Look at the photo above. Do you think the knee creases of this person are equal? Could either her right tibia or femur, or both, be longer than the left, causing her pelvis to be raised on the right?

STEP 6   Thigh Bulk

Compare the bulk of the client's left and right thighs. Are they equal?

What Your Findings Mean   Greater thigh bulk on one side suggests an increased use of the thigh muscles of that leg with respect to the other. An alternative explanation might be poor lymphatic drainage, as is seen in patients with lymphoedema. Considerably decreased bulk is observed in clients following illness or immobility and is due to muscle atrophy.

TIP   Clients who have injured a leg, foot or ankle are often observed to have less thigh bulk on that side simply because they are using that limb less. This may be accompanied by a compensatory increase in the bulk of the thigh on the other side. So a client recovering from a ruptured right Achilles tendon could have reduced bulk on the right lower limb and increased bulk in the left lower limb.

STEP 7   Genu Varum and Genu Valgum

Take a quick glance at the knees of your client, observing knee alignment and the overall shape of the joint.

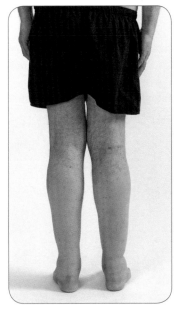
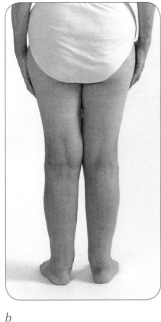
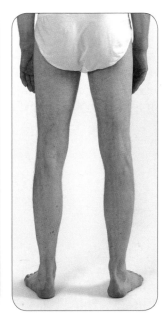

a   b   c

Photo *a* shows slight genu valgum of the right knee.

Photo *b* shows slight genu valgum of the left knee.

Photo *c* shows slight genu varum of the right knee. Observe the possible bowing in the right tibia of this person.

What Your Findings Mean   In some cases you will observe genu varum; in others, genu valgum. See step 5 of the lower body anterior postural assessment (on page 110) for more information.

STEP 8   Posterior Knees

Take a look at the posterior aspect of the knee, and note anything unusual about it. It is important to note whether a client stands with neutral, flexed or hyperextend knees; this is best done when you carry out the lateral postural assessment. However, you can sometimes get a feel for knee position by observing how prominent the popliteal area appears to be. Is there any oedema or signs of bursitis?

What Your Findings Mean   If the posterior knee seems more deeply creased than normal, this could indicate that the client is standing with a flexed knee. If the posterior knee is prominent, with the popliteus muscle seeming to protrude slightly, this could indicate that the client is hyperextending at this joint. Bursitis presents with an obvious protrusion.

STEP 9   Calf Bulk

Look now at the shape and bulk of your client's calf muscles. Are the calves even in girth, or does one appear more bulky than the other?

What Your Findings Mean   As in step 6 (thigh bulk), a larger calf muscle could indicate greater weight bearing or overuse on that side compared to the other side. A smaller calf suggests less use or atrophy, which is common following a prolonged illness or immobility.

TIP   You may observe that a client who fractured a leg or ankle as a child or teenager may have a smaller calf on the side of the former injury. Reduced weight bearing during childhood may affect muscle and bone development. Although subtle, you may observe some clients shifting their weight laterally with a subconscious disinclination to bear weight on the side of a former injury.

STEP 10   Calf Midline

Imagine a line running vertically down the centre of the client's calf from the knee crease to the Achilles tendon. If necessary, draw on this line using a body crayon. Compare the left and right calves and their relationship to the midline of the body.

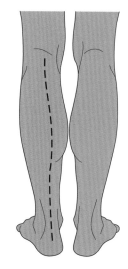

TIP One way to understand how hip rotation can affect the position of the calf is to draw the vertical calf lines on your client and then stand back and observe these lines when you instruct the client to alter her hip position. Ask her first to stand with one foot pigeon-toed. Compare the calf line on this leg with that of the other leg, and you will see that the line has moved outwards, away from the midline of the body, as the client has rotated the hip internally to stand pigeon-toed. Then ask her to turn her foot out on that side while keeping the other foot facing forwards or in a neutral position. This time the opposite happens: the calf line moves inwards, towards the midline of the body, as the client contracts the external hip rotators.

What Your Findings Mean The experiment described in the preceding tip box demonstrates that a line that appears to be lateral (rather than central) on the calf could result from an internally rotated hip on that side or a tibia that is medially rotated against the femur on that side. In either case, the foot position may also change when the person stands pigeon-toed. A line that appears to be medial (rather than central) on the calf indicates the opposite: An externally rotated hip on that side or a tibia that is laterally rotated against the femur. In this case, the client may stand with the feet turned out. Table 3.4 summarises this information and provides a reminder of the muscles acting on the hip to bring about either internal or external rotation.

If a client comes to you with a hip problem, a postural assessment is a good place to start because it may reveal shortness in one group of muscles and the need to test for tightness in these muscles later. Remember that it is ultimately important to discern whether the position of the calf is due to imbalances in hip muscles or torsion in the tibia because your treatment protocol will be different for each.

Table 3.4   Calf Line and Corresponding Effects on Feet and Leg Muscles

	Calf line appears lateral	Calf line appears medial
Hip or tibia position, or both	Indicates internal rotation of the hip, the tibia, or both	Indicates external rotation of the hip, the tibia, or both
Foot position	Sometimes the client stands pigeon-toed	Sometimes the client stands with the feet turned out
Muscles that may be shortened	Internal rotators of the hip: Gluteus minimus Gluteus medius (anterior fibers) Adductors Pectineus Gracilis	External rotators of the hip: Gluteus maximus Gluteus medius (posterior fibers) Piriformis Quadratus femoris Obturator Gemelli muscles Psoas* Sartorius

*The psoas is not a definitive rotator, yet recent research suggests it may be more involved in stability of the spine, including rotation, than originally thought.

STEP 11   Achilles Tendon

Take a look at the Achilles tendon and the position of the calcaneus. If necessary, draw a line vertically down the Achilles tendon, over the calcaneus and to the floor. Then stand back and observe the lines you have drawn. Is the tendon straight, concave, or convex? Do the feet appear to roll out or to roll in?

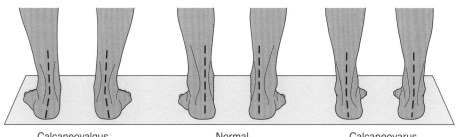

Calcaneovalgus   Normal   Calcaneovarus

Here are six ankles belonging to three clients. Observe the variety of shapes of the Achilles tendon, the position of the calcaneus, the position of the ankle joint itself, plus the foot position chosen by clients when undergoing postural assessment.

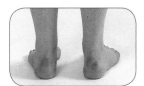  

What Your Findings Mean   The observation of the Achilles tendon can help provide information about excessive ankle eversion or inversion. Clients with excessive eversion, sometimes popularly referred to as overpronation, may have shortened peroneal (fibular) muscles on that leg.

STEP 12   Malleoli

When viewed posteriorly, the medial and lateral malleoli (of the same ankle) are not level. The medial malleolus is superior to the lateral malleolus. However, the lateral malleolus of the left ankle should be level with the lateral malleolus of the right ankle, and the medial malleolus of the left ankle should be level with the medial malleolus of the right ankle. The person in this photograph has prominent medial malleoli. Could it be that her tibiae are torsioned so that the knees turn inwards as the whole of the tibiae, including the distal end, rotate medially? Could that explain why we can see more of the medial malleoli on both the left and right ankles?

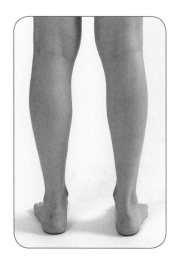

What Your Findings Mean   The figures here illustrate how the malleoli and calcaneal bones change position when a person has pes valgus or pes varus. In pes valgus, the medial malleolus appears superior to the position of the medial malleolus in the normal foot, and the lateral malleolus appears inferior to the lateral malleolus of the normal foot. The talus and calcaneus slope inwards, away from the midline of the leg, with ankle eversion. This indicates a weakness in the muscles that produce supination of the foot, including triceps surae, tibialis posterior, flexor hallucis longus, flexor digitorum longus and tibialis anterior. There is increased pressure through the medial side of the foot. People who stand with a pes valgus posture may have shorter fibular (peroneal) muscles.

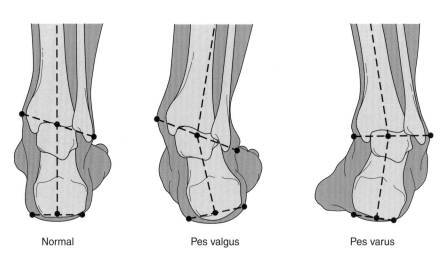

Normal   Pes valgus   Pes varus

In people who stand with pes varus, the calcaneus is inverted (supinated). The medial malleolus appears inferior to the position of the medial malleolus in the normal foot, and the lateral malleolus appears superior to the lateral malleolus of the normal foot. There is an alteration in the position of the talus and calcaneal bones as shown on page 64. This position correlates to a weakness in the pronators of the foot: fibularis, extensor digitorum longus and extensor hallucis longus. Table 3.5 summarises this information.

Table 3.5   Changes Corresponding to Pes Valgus and Pes Varus Foot Positions

	Pes valgus	Pes varus
Foot position	Everted (pronated)	Inverted (supinated)
Position of malleoli with respect to their position in a normal foot	The medial malleolus is superior; the lateral malleolus is inferior.	The medial malleolus is inferior; the lateral malleolus is superior.
Lengthened, possibly weak muscles	Supinators of the foot: triceps surae, tibialis posterior, flexor hallucis longus, flexor digitorum longus and tibialis anterior	Pronators of the foot: fibularis and extensor digitorum longus.
Weight bearing	More through the medial side of the foot	More through the lateral side of the foot

STEP 13   Foot Position

Finally, take a look at how the client has subconsciously positioned the feet. Each foot usually turns out equidistant from the midline of the body.

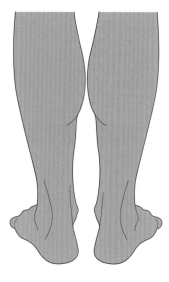

TIP   One way to assess foot position in the posterior view is to ask yourself how much of the lateral side of the foot you can see, how many toes. The more of the lateral aspect of the foot you can see (i.e., the more toes), the greater the degree of toe-out on that side.

What Your Findings Mean   As you learned in step 10, the position of the feet (and leg) ties in with the position of the hip and the tibia. Refer back to page 62 to see which muscles might be shortened in people who stand pigeon-toed and in those who stand with their feet turned out.

TIP   If, based on the position of her feet, you suspect that your client has shortened hip rotators, a crude but effective test is simply to ask her to stand in the foot position that would stretch those rotators. For example, if the client stands pigeon-toed, ask her to stand with her feet turned out like a ballet dancer. If the internal rotators really are tight, the client will find this toe-out position slightly uncomfortable.

STEP 14   Other Observations

Finally, as you did for the upper body posterior postural assessment, make note of any scars, blemishes or unusual marks on the client's skin. Has any strapping or taping been applied perhaps in the treatment of an injury?

Quick Questions

1. Which muscles laterally flex the neck to the right?
2. When observing muscle bulk, you notice atrophy of the shoulder muscles. What might account for this?
3. What does *winging of the scapula* mean?
4. Which low back muscle might be shortened in a client laterally flexed to the left with left elevation of the pelvis?
5. What are two reasons the midline of the calf might be more lateral on one leg than on the other leg?

Lateral Postural Assessment

Now that you have carried out at least one posterior postural assessment, it's time to work through similar steps, this time viewing your client from the side. As in chapter 3, here you will learn what to look for as you compartmentalise the body with this step-by-step approach. Of course, in reality, the various parts of our bodies do not function in isolation, so it is important to finish this section of the assessment by taking in an overall view of your client, as indicated on page 90. Start by locating the lateral postural assessment chart on page 146. There are 15 steps—eight for the upper body, six for the lower body, and one in which you look at your client's overall posture. You will need to compare left and right sides of the body. To save time (and a lot of moving about), work through each of the steps in turn, examining one side of the body before turning to the other.

STEP 1   Head Position

Start by assessing the position of the client's head relative to the body. Does the head appear to sit comfortably over the thorax? Or does it appear to be pushed forwards, chin out, as if the person were rushing?

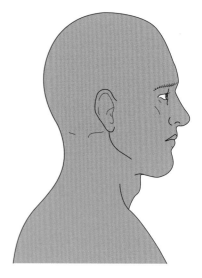

What Your Findings Mean   A forward head posture affects the neck, chest and arms. It is important to recognise that in this posture the lordotic curve of the neck is not necessarily increased; rather, the head is positioned too far in advance of the body. Cervical extensor muscles such as the levator scapulae are therefore *not* shortened and tight as many therapists believe, but are lengthened and weak. Theoretically, such a posture increases the strain placed on posterior cervical soft tissues and may result in neck, shoulder and upper back pain.

TIP   Think of the levator scapulae muscles as being the reins of a horse constantly pulling in the head, bringing it back over the centre of the body. As the head moves ever more anteriorly, the reins (muscles) are ever more lengthened.

STEP 2   Cervical Spine

Next, take a look at the client's cervical spine. If the client's hair obscures her neck, either have her tie up her hair, or leave this section of your assessment blank. How does the cervical spine look? Does it have the normal lordotic curve, or is the curve exaggerated? Rarely, clients have what appears to be a flatter curve than is normal. Is your client like that?

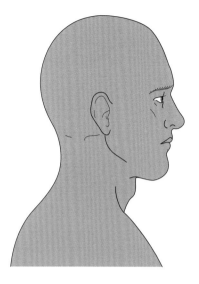

TIP An exaggerated lordotic curve in the cervical spine often accompanies a kyphotic posture. To understand why this might be, try this: sitting, allow yourself to slump, exaggerating the kyphotic curve in the spine associated with poor posture. Notice that your eyes naturally look downwards as your head falls forwards. Maintaining this slumped position, look up, imagining a computer screen in front of you. The lordotic curve of your neck has now increased as your face is raised to look forwards.

What Your Findings Mean   An increase in the normal curvature of the cervical spine possibly increases compression on the posterior part of some of the cervical intervertebral discs. The zygapophyseal joints may be compressed, too. Because this posture is also associated with an exaggerated kyphosis in the thoracic spine, the thoracic cavity may be diminished. A reduced thoracic cavity is associated with shortened intercostals, pectoralis minor, adductors and internal rotators of the shoulders. Muscles that are often weak in a kyphotic posture include the thoracic spine extensors and the middle and lower fibers of the trapezius.

Exaggerate the lordotic curve of a model spine, and you can see that the bifurcated spinous processes of the cervical vertebrae begin to approximate each other. Structurally, the cervical extensor muscles are brought closer together and are therefore likely to be shortened and weak, and the neck flexor muscles are likely to be lengthened and weak. If a person maintains this posture over many years, it seems reasonable to assume that adhesions would form between joint capsules and surrounding structures resulting in a decrease in range of movement. Could prolonged compression of some of the cervical vertebrae even result in the development of osteophytes in this region of the spine? Conversely, if the cervical spine looks unusually flat, this indicates shortness in the neck flexors and weakness in the neck extensors.

STEP 3   Cervicothoracic Junction

Turn your attention now to the junction between vertebrae C7 and T1. Does it look normal, or is there an increase in soft tissue in this area, or a hump shape?

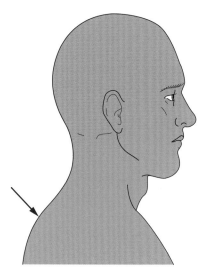

TIP   To make the C7 vertebra easier to locate when viewing your client laterally, simply ask him to flex his neck, thus making the spinous process more prominent.

What Your Findings Mean   A dowager's hump describes this raised area in the C7/T1 junction often observed in postmenopausal women where osteoporotic changes cause vertebrae to become wedge shaped anteriorly. Could the fatty tissue deposited over the C7/T1 junction of some clients be the result of poor posture?

Now look at the photographs of these eight clients. From what you have read in steps 1, 2 and 3, are you able to identify which of these clients has a forward head posture? Who has an increased lordotic cervical curve? Do any appear to have elongated necks? Compare the cervicothoracic junction on each.

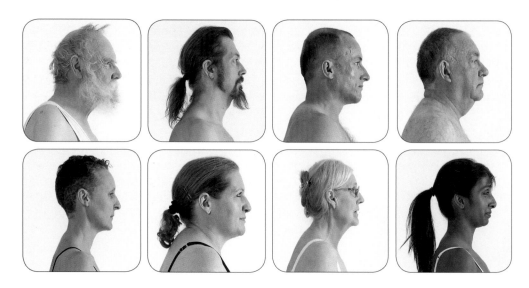

STEP 4   Shoulder Position

Look at the shoulder nearer to you. What is the relationship of that shoulder to the head and neck? Does it sit nicely in line with the ear? Or does it appear protracted, the arm falling into internal rotation? Alternatively, does the client stand erect with the chest out and the shoulders pulled back in a military-style posture? Take a look at the photographs of the people in step 6 on page 77. If you had to select the person with the most internally rotated right shoulder, whom would you pick?

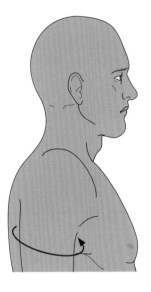

TIP   One way to appreciate the connection between poor posture and rotation of the humerus is to try this exercise. Standing, notice the position of your hands and arms. Now deliberately slump, giving yourself poor posture. Notice what has happened to your hands. Do your thumbs and the radial sides of your wrists touch your hips? By comparison, notice what happens when you stand with your chest up and shoulders back. You may find that as the scapulae retract and your humeri are drawn laterally, and with it your forearms, your thumbs no longer touch your thighs.

What Your Findings Mean   Protraction of the shoulders is one of the most common postures you will encounter because many people sit for long periods at desks or while driving, their arms in front of them at a keypad or steering wheel. With time, this slumped position becomes habitual, with consequences for the chest and neck as well as the shoulder joint itself. This position is associated with lengthened and weak rhomboids, tight pectorals and shortened intercostals. Conversely, the middle and lower fibers of the trapezius may be lengthened and weak also, as might be the extensors of the thoracic spine. An internally rotated humerus suggests shortness in the muscles of medial rotation. Could such a position contribute to shoulder impingement syndromes?

Retracted shoulders are less common than protracted shoulders and are associated with the military-style posture. Here there may be shortness in the rhomboid muscles and in the middle fibers of the trapezius, and some parts of the pectoralis major may be lengthened. An externally rotated humerus suggests shortness in muscles such as the infraspinatus and teres minor. It is quite possible to have a protracted shoulder on one side of the body, with the accompanying internal rotation on that side, and a retracted shoulder on the other side of the body, with accompanying external rotation. One cause of this might be regularly wheeling a heavy luggage trolley behind on one side, externally rotating on that side.

TIP To understand how a protracted shoulder on one side might come about, imagine that you are pulling a heavy wheeled trolley in one hand. Notice that on the side of the trolley, your scapula is retracted and the humerus is externally rotated. Notice how you are also supinated at the wrist and elbow and that the other side (the side without the trolley) becomes the leading side, with the scapula protracted and an internal rotation of the humerus.

STEP 5   Thorax

The lateral postural assessment is a good opportunity to look for exaggerations in the thoracic curve, commonly seen in clients who habitually adopt poor posture while sitting. Older adults, especially those who are sedentary, develop a similar posture, which is due in part to age-related changes in vertebrae. An exaggeration in the thoracic curve may be compensatory, accompanying an increase in the cervical or lumbar lordosis, or both. By complete contrast, in some clients the normal curve in this region is markedly reduced, giving the client the appearance of a flat back.

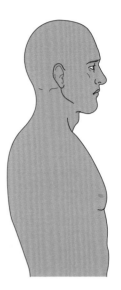

TIP   In a client with very low body fat or atrophied muscles, spinous processes may appear more prominent than normal. It is a mistake to assume that because you can see these processes, the client is kyphotic. What you may be observing is simply a normal bony structure made more apparent by the decreased covering of body tissue.

What Your Findings Mean   A severely kyphotic posture is associated with shortened pectorals, tight intercostals, and perhaps even shallow breathing due to a depressed chest cavity. There may also be shortening of the upper abdominals. Thoracic spine extensors, the middle and lower fibers of the trapezius, and the rhomboids may be lengthened and weak. A person with a kyphotic posture associated with poor sitting may exhibit internal rotation of the humerus with the accompanying changes in the length and strength of the SITS muscles (supraspinatus, infraspinatus, teres minor and subscapularis). Not surprisingly, people with kyphotic postures often have neck and shoulder pain.

A decreased kyphotic curve is sometimes observed in clients with increased flexibility or hypermobilty syndromes, in which the thoracic region appears flatter than usual. These clients often complain of thoracic pain, perhaps because the spinous processes of the vertebrae in this region start to approximate each other when the clients sit erect or stand upright.

STEP 6   Abdomen

An area that sometimes gets overlooked in postural assessment is the abdomen. How does the abdomen of your client appear? Is it flat or protruding? In a normal, healthy person, the abdomen should be flat.

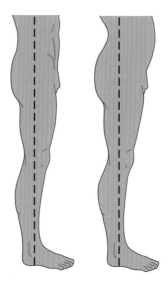

Take six abdomens! The photographs on the opposite page demonstrate the variety in the shape and position of the abdomen when a person is viewed laterally. Does an abdomen protrude because the person is overweight or pregnant, or it is the result of the person's overall standing posture and an anteriorly tilted pelvis? Is there increased tension in the abdomen perhaps corresponding to a posteriorly tilted pelvis and a decreased curve in the lumbar spine?

What Your Findings Mean   Protrusion of the abdomen could be a natural consequence of pregnancy or the result of increased lumbar lordosis, or it could simply be excess adipose tissue because the client is overweight. Clients with restrictions in the muscles and fascia of the chest sometimes appear to have a protruding abdomen, quite a distinct change in shape from the chest area, which is tight and depressed.

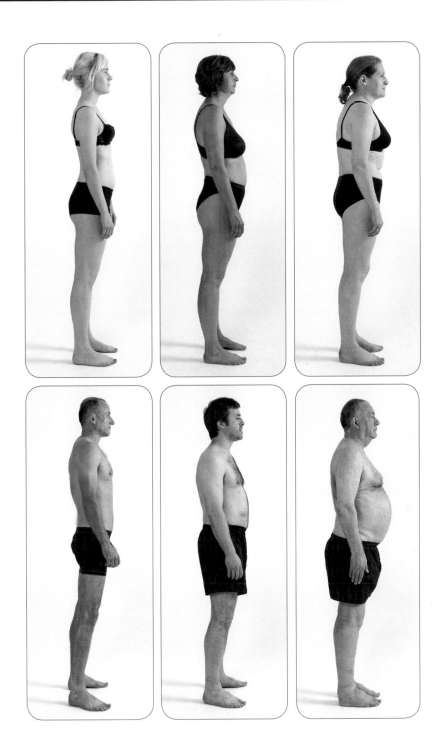

STEP 7   Lumbar Spine

This is a good point at which to take stock of your findings. You may have noticed that nowhere in the text has it been suggested that you *should* treat a client based on a particular observation of posture. Looking at the lumbar spine, it is useful to remember that some people have naturally increased lumbar curves and are asymptomatic; others may have mildly increased lumbar and thoracic curves and require no treatment at all.

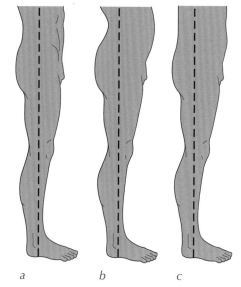

a   b   c

The lumbar spine and pelvis are inherently linked. Increases or decreases in the lumbar curve correspond with altered positions of the pelvis. However, it can be difficult to identify pelvic position when you are first learning postural assessment, so observing the lumbar region first is a good place to start. Does the curve of your client's lumbar region look normal, or is there evidence of increased or decreased lordosis? The figures on page 79 might help you understand where the lumbar vertebrae lie with relation to a plumb line when a client has a normal (a), increased (b) or decreased (c) lumbar lordosis.

What Your Findings Mean   An increased lordotic curve indicates an anteriorly tilted pelvis. For a full explanation of pelvic positions, please see page 76. Increased lordosis in this region could explain pain resulting from compression of soft tissues in this region—for example, if there were increased compression on the posterior part of the lumbar intervertebral discs and shortening of the lumbar erector spinae. The correct functioning of the zygapophyseal joints in the lumbar spine may also be compromised if these too are compressed. The rectus abdominis may be longer and therefore weaker than usual, as may the hip extensors, whereas the muscles responsible for lumbar extension may be shortened.

The next time you are treating a client with an increased lordotic curve, you might find it helpful to assess the length of the hamstrings. Anatomically, a lordotic curve accompanies an anteriorly tilted pelvis, and therefore, the hamstrings are held in a lengthened position. Yet in my experience, such clients often complain of tight hamstrings. Could this be because their hamstrings are trying to pull the ischium back into a more neutral position? The psoas may also be shortened if it is responsible for pulling the bodies of the lumbar vertebrae anteriorly, contributing to lordosis in this region of the spine. An interesting question: Which came first, the shortened psoas or the increased lordosis?

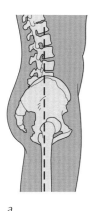
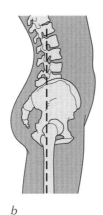
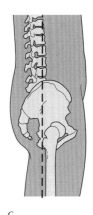

a   b   c

Anatomically, an obvious decrease in the normal curve of the lumbar spine may be associated with tightness in the hip extensors and longer, and perhaps weakened, hip flexor muscles. Table 4.1 summarises this information.

Table 4.1   Factors Associated With Changes in the Lumbar Curve

	Increased lumbar lordosis	Decreased lumbar lordosis
Corresponding position of pelvis	Anteriorly tilted	Posteriorly tilted
Shortened muscles	Extensors of the lumbar spine	Hip extensors
Lengthened muscles	Rectus abdominis Hip extensors	Hip flexors

The client shown here is a good example of someone with an increased lordosis in the lumbar spine. If you look back in this book to all of the previous photographs showing full-length lateral views, you will observe that in most cases you can see the curve of the lumbar region.

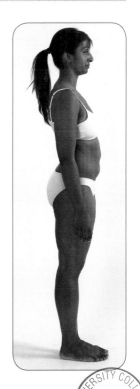

STEP 8   Other Observations

Use this last step to note any scars, blemishes, discolouration or swelling or anything else not yet documented in your upper body assessment.

STEP 1   Pelvis

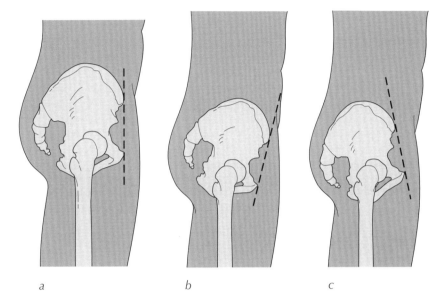

a   b   c

As you know, movements of the pelvis correspond to changes in the shape of the lumbar spine. A few of the ways it can move is to tilt anteriorly (*b*) or posteriorly (*c*) away from the normal (*a*) position. *Anterior pelvic tilt* describes the position of the pelvis when the anterior superior iliac spines (ASIS) are positioned anterior to the pubis. *Posterior pelvic tilt* describes the position of the pelvis when the ASIS are positioned posterior to the pubis.

TIP To get your head around anterior and posterior pelvic tilting, try this: Standing, push your abdomen forwards and your buttocks out, extending your lumbar region. This produces an anterior pelvic tilt. Return to your neutral, resting position. Now contract your buttocks, pushing your groin forwards and flattening your lumbar spine. This produces a posterior pelvic tilt.

TIP There is a trick to help you determine whether your client is standing with a particularly lordotic lumbar region, or whether this region is flattened. Ask your client to perform the tilting maneuvers you tried for yourself in the preceding tip. Once she understands what to do, observe what occurs as she performs an anterior tilt, and then a posterior tilt. If the client has difficulty tilting her pelvis anteriorly, increasing her lumbar curve, this could be because she is already in an anteriorly tilted position. If she has difficulty posteriorly tilting her pelvis, flattening the lumbar curve, this could be because she is already in a posteriorly tilted position.

What Your Findings Mean   When the pelvis tilts anteriorly, the curve of the lumbar spine becomes exaggerated so your client may appear lordotic in this region. There may be increased compression on the posterior part of the lumbar intervertebral discs, plus compression of the zygapophyseal joints in this region. This posture is associated with lengthened but weak hamstrings, plus a lengthened rectus abdominis. The psoas may be shortened, and there may be shortened rectus femoris, too. With the pelvis posteriorly tilted, the lumbar lordosis is decreased. This position is associated with tightness in the hip extensors and longer, and perhaps weakened, hip flexor muscles. Table 4.2 summarises this information.

Table 4.2   Factors Corresponding to Anterior and Posterior Pelvic Tilt

	Anterior pelvic tilt	Posterior pelvic tilt
Position of the ASIS	The ASIS are held anterior to the pubis	The ASIS are held posterior to the pubis
Corresponding position of the lumbar spine	Increased lordosis	Decreased lordosis
Shortened muscles	Extensors of the lumbar spine	Hip extensors
Lengthened muscles	Rectus abdominis Hip extensors	Hip flexors

Remember also that other areas of the spine are likely to change shape to compensate for the position of the pelvis.

STEP 2   Muscle Bulk

Next, observe the muscles of the lower limbs paying particular attention to the thighs and gluteus maximus. Is there an increase or a decrease in muscle bulk between the left and right sides of the body?

What Your Findings Mean   Muscles atrophy with disuse. This is commonly observed in sedentary elderly clients and in younger clients who have had all or part of a lower limb immobilised, usually following injury. The longer the period of immobility, the greater the degree of atrophy. A decrease in muscle bulk might also be observed in clients who are unable to fully bear weight on that leg, perhaps with an increase in muscle bulk on the opposite leg resulting from increased weight bearing on that side.

STEP 3   Knees

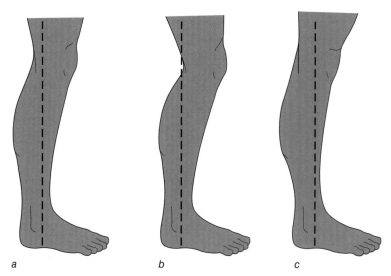

a   b   c

The lateral assessment is an excellent opportunity to observe what is happening at your client's knee joint. Are the knees normal (*a*), flexed (*b*) or hyperextended (*c*)?

TIP If you can see more of the popliteal area (and perhaps the calf, too) of the right leg when viewing the left side of the client, this indicates that the right knee is hyperextended. Being able to see more of the left leg when viewing the client's right side suggests that there may be increased extension in the left knee joint.

The person in the photo on the left is a good example of someone who stands with increased extension at her knee joint. Observe the front of this woman's knee. Can you see how it appears to be compressed, the patella pushed into the front of the joint? Can you see how if you were to draw a plumb line onto this image, the leg would fall posterior to the plumb line? (Remember that when using a plumb line you position your line just anterior to the lateral malleolus). The person in the photo on the right has a similar, but less obvious knee position.

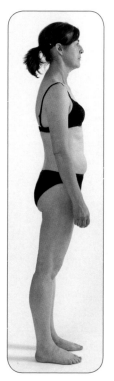

By contrast, take a look at these two photographs. The first person is standing with flexed knees. This is less obvious than usual because the lower limbs are swollen. The second person is standing with a flexed right knee. This is subtle but may be easier to see once you have compared it to the other images of the lateral posture in this book.

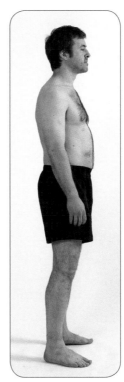

What Your Findings Mean Flexed knees are associated with tight hamstrings and popliteus muscles and weak quadriceps and soleus muscles. Just as the position of the pelvis and lumbar spine are intimately related, and changes in the position of one correspond to changes in the position of the other, knee position affects the hip and ankle joints. A flexed knee may accompany an increase in flexion at the hip and an increase in dorsiflexion at the ankle joint. You can demonstrate this yourself by simply altering the position of your knee in standing, from neutral to flexed.

Certain pathologies affecting the knee inhibit extension of the joint. For example, a loose body within the joint may prevent full extension; the pain of chondromalacia patella may be aggravated by full extension. Clients who are hypermobile often hyperextend their knees in standing unless they have learned to avoid this. Hyperextended knees are associated with tight quadriceps and lengthened hamstrings. Tight quadriceps in a client with hyperextended knees may contribute to anterior knee pain as the patella is pushed against the femur in standing. Could this posture contribute to degenerative changes in the cartilaginous surfaces of the patellofemoral joint? Another consequence might be increased stress on the posterior aspect of the joint capsule. Hyperextended knees are also associated with decreased dorsiflexion at the ankle joint. Table 4.3 summarises this information.

Table 4.3   Factors Corresponding to Changes in Knee Joint Position

	Flexed knees	Hyperextended knees
Shortened muscles	Hamstrings Popliteus	Quadriceps
Lengthened muscles	Quadriceps Soleus	Gastrocnemius
Hip position	Increased hip flexion	Increased hip extension
Ankle position	Increased dorsiflexion	Decreased dorsiflexion
Other	Increased pressure on structures of the anterior ankle joint	Stretching of the posterior joint capsule of the knee; increased likelihood of degenerative changes to the patellofemoral joint

STEP 4   Ankles

a   b   c

Moving away from the knees now, examine your client's ankles. Are they neutral (a), or do you notice any increased (b) or decreased (c) dorsiflexion? The three people here are good examples of decreased dorsiflexion at the ankle.

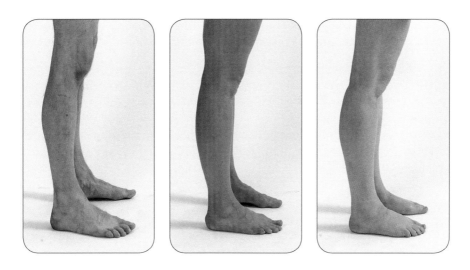

What Your Findings Mean   Increased dorsiflexion in standing is observed in clients who stand with flexed knees. In these clients, ground forces are no longer distributed evenly up through the tibiae during walking. One might postulate that the consequence of this might be pain and early degenerative joint changes. There may be a shortened tibialis anterior muscle and increased pressure to the anterior aspect of the ankle retinaculum. Decreased dorsiflexion is associated with shortened quadriceps and increased pressure to the anterior of the knee joint.

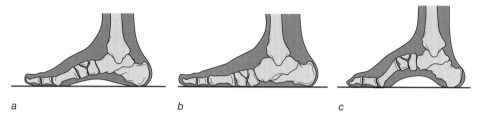

STEP 5   Feet

a   b   c

Finally, how do the foot arches appear to you? Are they normal (a), dropped (b) or elevated (c)? Does your client appear to bear weight equally between the left and right feet, or is he putting more weight on one foot than the other? Are there any marks on the feet or ankles? Do the toes appear normal, or is there evidence of claw toes or hammer toes? This photo illustrates a good example of pes planus.

What Your Findings Mean   Marks on the skin of the feet suggest that footwear or supportive aids are too tight, causing compression of tissues or rubbing of the skin. Toe problems may explain why some clients have problems with balance, especially when these problems affect the first toe.

Increasing pressure on the lateral side of the left foot corresponds with a trunk rotated to the left; increased pressure to the lateral side of the right foot corresponds to the trunk rotated to the right.

TIP   You can demonstrate how rotation in the trunk affects feet in the following way. Standing in bare feet in your normal stance, rotate your trunk one way as far as you can while keeping the soles of your feet on the ground. Notice how the points of pressure change between your plantar surface and the floor.

However, rotation of the trunk may *not* be the cause of increased pressure on one side of the foot in all clients; it could be due to some other biomechanical factor. Understanding the biomechanics of feet and ankles is a special area of bodywork. If you think that problems in this area might be contributing to your client's condition, it might be worth considering referring your client to a podiatrist for an assessment.

STEP 6   Other Observations

Use this step to make any observations not yet recorded in the earlier steps for the lateral view of the lower half of the body, such as scars and bruising. For example, the person in this photograph has oedematous feet. Observe also the second toe of this man's right foot.

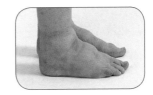

To complete your lateral assessment, stand back and take an overall view of your client. The illustrations provided here represent 'classic' postures. The second shows a person with a kyphotic thoracic spine and the associated increased lordotic cervical and lumbar curves. The third shows a client with a flattened lumbar region. The fourth shows a sway-backed posture.

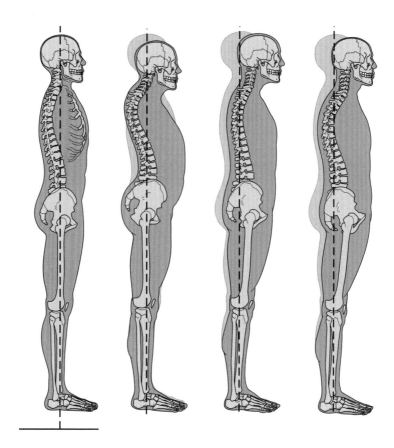

Look at these four photographs. If you were to draw a vertical plumb line up from each person's ankles, would you be able to match the postures of any of these people to the postures illustrated in the previous figures? Perhaps you can match the lower half of one photograph to the lower half of one figure, but the top half of the photograph to the top half of a different figure? Or, perhaps in reality, we need a far wider range of postural types to which we can match ourselves?

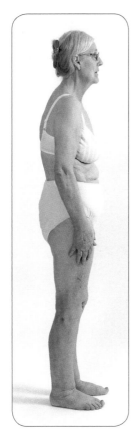

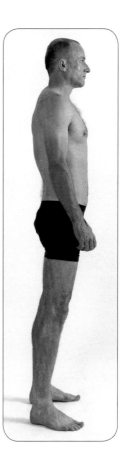

Quick Questions

1. What are the consequences of having a forward head posture?

2. Which muscles are shortened in an internally rotated humerus?

3. Which sorts of activities might increase the kyphotic curve of the thorax?

4. When the pelvis tilts anteriorly, does the lordotic curve of the lumbar spine increase or decrease?

5. Is a client who stands with flexed knees likely to have shortened hamstrings or shortened quadriceps?

Anterior Postural Assessment

The final part of your standing postural assessment is to observe your client anteriorly. Many experienced therapists choose to view clients from the anterior position at the start of their assessments, before they move on to the posterior and lateral views. However, when you are first learning, it might be better to leave the anterior view until last because some participants feel intimidated when observed this way, especially if you are taking more time over the assessment. The anterior postural assessment chart can be found on page 149. As with the previous two chapters, use this to document your observations as you work through each of the steps presented here. There are 25 steps: 11 for the upper body, 13 for the lower body and one to finish that requires you to take an overall view of your client.

STEP 1   Face

Without being too invasive, observe your
client's face and note any asymmetry.
What does the face tell you about the
health of your client? Does he appear to
have good skin tone and be healthy and
nourished, or is he sallow and pale? Does
he look alert or tired? Is the face drawn
or bloated? Much can be learned from
facial expressions. Does your client look
as though he is in pain? Relaxed? Worried?
Look also at the tone of facial muscles. Is
the client involuntarily clenching his jaw,
for example, or frowning? Is there any
spasming of facial muscles?

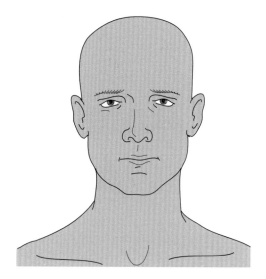

What Your Findings Mean   Although textbooks suggest that we are all symmetrical, in
reality this is not the case. It is quite normal to have variations in facial features just as
it is to have variations in other parts of the anatomy. However, spasming or flaccidity
in muscles is not normal, and these should be noted.

STEP 2   Head Position

Is the head positioned so that the nose falls in the midline along with the manubrium, sternum and umbilicus? Or is there any lateral deviation or rotation away from the midline?

What Your Findings Mean   There are many reasons for asymmetry in the head and neck. For example, rotation or lateral flexion to a minor degree is often observed in people who spend prolonged periods in a fixed position with their workstation to one side of them rather than directly in front of them. Severe lateral flexion with or without rotation, combined with heightened tone in the sternocleidomastoid, could indicate torticollis. An altered head position may indicate that the client has suffered an injury to the neck.

This photograph shows a good example of why it is important to consider the interconnectedness of the body. If you were to draw a plumb line between the man's medial malleoli, between his knees and through his umbilicus, where does the position of this man's head fall with respect to the line? Is he laterally tilting his head to his right, shifting his body weight to his right or both?

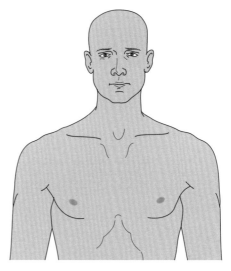

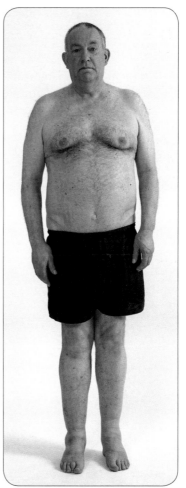

STEP 3   Muscle Tone

Do any of the muscles of the neck, chest and shoulders appear more prominent on one side of the body than on the other? Pay particular attention to the sternocleidomastoid, the scalenes and the upper fibers of the trapezius. Conversely, is there any decreased tone or atrophy?

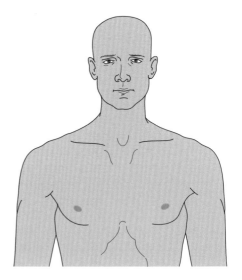

What Your Findings Mean   A muscle that appears more prominent suggests an increase in tone in that muscle, which, if prolonged, could contribute to pain in that area. This raises the question, What is the client doing to increase tone in that muscle? Hypertrophy of the pectorals (and many other muscles) is commonly observed in bodybuilders. Increased tone in respiratory muscles (such as the scalenes and the sternocleidomastoid) is observed in many people. Scalenes might appear particularly prominent in people with long-term respiratory conditions such as chronic obstructive pulmonary disease. Atrophy, on the other hand, indicates disuse. You might observe atrophied neck muscles when assessing a client who has been immobilised in a collar after cervical trauma.

STEP 4   Clavicles

Observe both the angle and contour of the clavicles. Both should have smooth contours and should be gently angled upwards away from the sternoclavicular joint. Look also at the acromioclavicular (AC) joint.

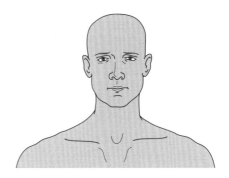

TIP   Standing facing a mirror, observe your clavicles. Look at the angle they form with respect to the sternum as they slope gently upwards towards the AC joint. Now shrug, elevating your shoulders. Can you see how the angle of the clavicle changes and becomes steeper?

To understand how movements of the scapula affect the position of the clavicle, the illustration here shows a posterior view of these two bones. The dashed lines show the position of the scapula as it rotates upwardly. Notice what has happened to the clavicle. As you can see, it has started to rise.

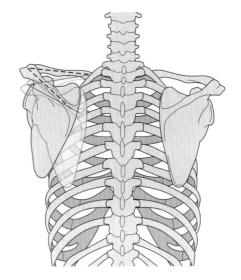

What Your Findings Mean   Sharply angled clavicles indicate elevated shoulders. It is normal for the clavicle on the dominant side to be lower than that on the non-dominant side. Uneven contours could indicate a fracture that has healed in mal-alignment, or a more recent injury such as a ruptured AC joint.

STEP 5   Shoulder Level

Are the shoulders approximately level and the contours of the deltoid muscle even on the left and right sides of the body?

What Your Findings Mean   It is common for the shoulder of the dominant hand to be slightly lower than the other. A client may elevate a shoulder to protect an injured or painful joint in the shoulder or in the neck. Depression of the shoulder plus indentation in the contour of the deltoid is observed in people with subluxation at the glenohumeral joint.

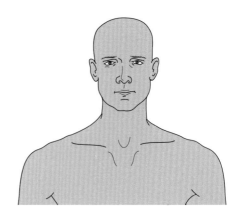

   Now that you have a greater understanding of postural assessment, look at the following photographs, which illustrate some of the steps you have covered in this chapter. Can you see how the upper fibers of the trapezius on the people in the first two photographs differ between the left and right sides? The left clavicle of the person in the third photo is raised.

STEP 6   Rounded Shoulders

Rounded shoulders are easier to identify from a lateral view. However, when the shoulders are internally rotated in this way, the position of the hands change, which you can observe during both your anterior and posterior assessments. If a client is internally rotated at the humerus, you may see more of the dorsal surface of the hand during the anterior assessment. The woman in the first photograph demonstrates this. Compare the positions of the left and right hands of the person in the second photo. Would you say her left shoulder is more internally rotated than her right?

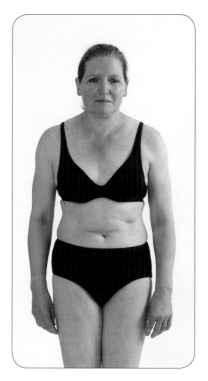 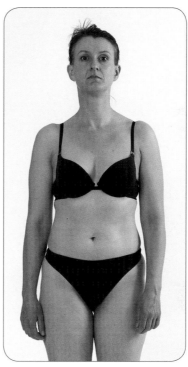

What Your Findings Mean   Rounded shoulders are associated with kyphotic postures and indicate tightness in the anterior chest muscles plus the internal rotators of the humerus. Could having an internally rotated humerus contribute to the impingement of the anterior shoulder structures such as the long head of the biceps brachii?

STEP 7   Chest

The thorax may shift laterally or rotate relative to the neck and pelvis. In this illustration, the spine is represented by a shaded line and the plumb line is represented by a dashed line. You can see that the client has shifted the thorax and head to the right.

To check for shifts in the thorax, ask yourself questions such as, Does the sternum appear in the midline? What about the rib cage—does it sit squarely over the pelvis? Does the rib cage appear rotated or shifted to one side?

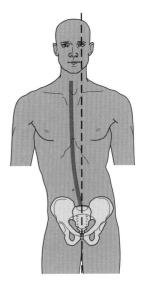

What Your Findings Mean   Shifts of the chest occur for many reasons. For example, the illustration above represents the kind of thoracic shift that is sometimes observed in a client with sciatica. However, that does not mean that if you observe this posture the client *is* suffering from sciatica. Lateral curvatures in the spine as well as muscle imbalances can also contribute to this posture. When the thorax rotates, compensatory changes occur in the neck and lumbar spine.

TIP   You can observe the effect rotation of the thorax has on the neck by rotating your chest to the right while keeping your head, neck and hips in the same position, facing forwards, while standing. Notice where you experience an increase in tension. You will find that to keep your head facing forwards as you rotate your trunk to the right, you need to contract the muscles of your neck that rotate your head to the left.

STEP 8   Carrying Angle

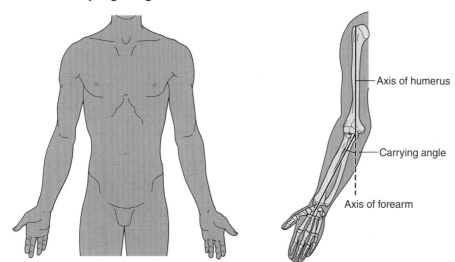

Axis of humerus

Carrying angle

Axis of forearm

The carrying angle is the angle formed between the long axis of the humerus and the long axis of the forearm. Ask your client to stand so she is in the anatomical position, with the palms of her hands facing forwards. Have her keep her elbow extended and her forearm supinated. What sort of angle does her elbow form?

What Your Findings Mean   In males, a normal angle is 5 degrees; in females, a normal angle is 10 to 15 degrees (Levangie and Norkin 2001). A carrying angle much greater or smaller than the norm is sometimes present following an elbow fracture. An abnormal carrying angle can affect a person's ability to bear weight through the upper limb (as when doing press-ups, for example).

STEP 9   Arms

Observe the form and bulk of the arm and hand muscles, comparing the left and right sides just as you did when examining the girth of the lower limbs during the posterior postural assessment. Notice, too, how the client positions his upper limb. Does he keep it close to his body, or does he allow it to hang loose in a more relaxed posture?

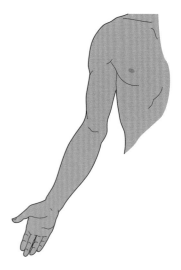

What Your Findings Mean   Increased bulk indicates increased use of that limb; decreased bulk or even atrophy indicates disuse and is a common observation in clients following immobilisation of the elbow, wrist or hand and even the shoulder. Increased abduction of the arm ties in with step 14, upper limb position, in the posterior postural assessment on page 46. An arm held close to the side of the body, or even across the body, indicates protectiveness of that limb.

STEP 10   Hands and Wrists

If you are assessing a client for a problem in the hands or wrists, it is obviously important to observe the hands in detail. A detailed analysis of hands and wrists is usually best performed with the client in a sitting position. However, when a client is standing, it is worth noting any abnormalities such as swellings, bruising or discolouration. Notice also any obvious changes in the position of the fingers, and pay particular attention to the thumbs and whether there is any wasting in muscles of the thenar or hypothenar eminence. Notice whether the wrist joints themselves are level.

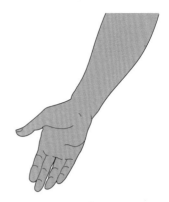

What Your Findings Mean   Many factors affect the appearance of the hands, fingers and wrists. For example, conditions such as rheumatoid arthritis are revealed by swollen, inflamed and often misshapen joints in the fingers. Obvious muscle wasting may be due to nerve damage or impairment. Discolouration can indicate poor blood flow to the extremities, which is common in conditions such as diabetes.

STEP 11 Abdomen

Observe the umbilicus. Does it lie in the midline along with the sternum and pubic symphysis? Are there any obvious surgical scars as in this photograph?

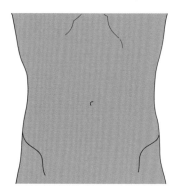

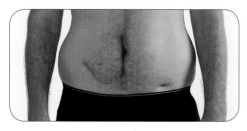

What Your Findings Mean   An umbilicus that does not fall in the midline ties in with rotation of the thorax and pelvis. Some therapists believe that rotation of the umbilicus to the right indicates a shortening in the iliopsoas muscles on the left, and that rotation to the left results in shortening of the iliopsoas muscles on the right. However, as with each of the steps in your postural assessment, it is important to keep an open mind as to causal factors.

STEP 1   Lateral Pelvis

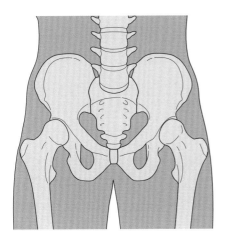

a

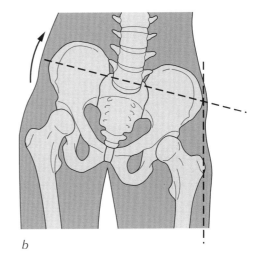

b

The anterior superior iliac spines (ASIS) of the pelvis should be level. Are they? Or is there any lateral tilt, indicated by one being lower than the other? The illustrations here show a normal pelvis (a) and the hip hitched to the right side (b).

What Your Findings Mean   As you learned in step 2 (pelvic rim) of the lower body posterior postural assessment, a lateral tilt in the pelvis corresponds with a lateral curve in the lumbar spine. When a client is hip hitched on the right, with a higher ASIS on that side, the right quadratus lumborum muscle could be shorter than the left. The right hip will be adducted (with corresponding shortening of the right adductor muscles), whereas the left hip is in an abducted position (with corresponding shortening in muscles such as the gluteus medius on that side).

STEP 2   Rotated Pelvis

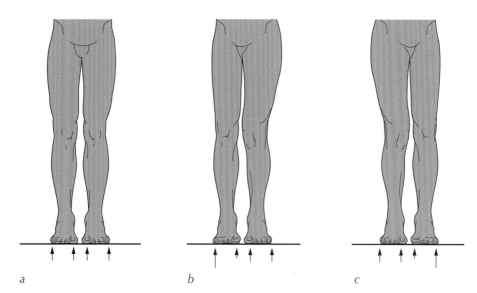

a   b   c

In the posterior postural assessment, you examined the pelvis to assess whether it was rotated. You can do the same here.

a: Normal pelvis with both ASIS aligned. Knees face forwards. There is equal pressure beneath the medial and lateral sides of the foot.

b: The whole pelvis is rotated to the right. Knees no longer face forwards. There is increased pressure on the lateral side of the right foot.

c: The whole pelvis is rotated to the left. Knees no longer face forwards. There is increased pressure on the lateral side of the left foot.

What Your Findings Mean   Rotation of the pelvis affects the feet and knees, as outlined in table 5.1. Rotation of the pelvis also affects the thorax. For a summary of this, please see page 55 in chapter 3.

Table 5.1   Pelvic Rotation and Its Effect on the Feet

ROTATION OF THE PELVIS TO THE LEFT	
Left foot	**Right foot**
<ul><li>Increased supination</li><li>There is increased pressure on the lateral side of the foot, and decreased pressure on the medial side of the foot as a result of increased inversion of the forefoot.</li></ul>	<ul><li>Increased pronation</li><li>The pressure on the lateral and medial sides of the foot are roughly equal.</li></ul>
ROTATION OF THE PELVIS TO THE RIGHT	
Left foot	**Right foot**
<ul><li>Increased pronation</li><li>The pressure on the lateral and medial sides of the foot are roughly equal.</li></ul>	<ul><li>Increased supination</li><li>There is increased pressure on the lateral side of the foot, and decreased pressure on the medial side of the foot as a result of increased inversion of the forefoot.</li></ul>

TIP   You can easily test the effect pelvic rotation has on the feet by rotating to the left and right and feeling what happens to the contact points of the soles of your feet and the floor.

Now that you have learned something about the pelvis, take a look at this photograph. Can you see that this person's pelvis is higher on the left side?

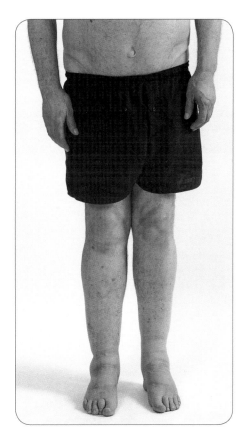

STEP 3   Stance

How does your client stand? Does she bear weigh equally through both limbs, or does she seem to favour one side? Does she naturally stand with her legs together, or has she chosen a wide stance?

What Your Findings Mean   Clients who stand in a wide stance create a wide base of support for themselves. Why might they do this? Is it because they feel unbalanced? Could it be that in some cases they have weak hip adductor muscles relative to their abductor muscles?

TIP If you notice that your client is standing in a wide stance, and providing you believe it is safe to do so, ask her to stand with her feet together (so that the medial malleoli of the ankles are as close as possible) and ask how she feels. Clients with weak adductor muscles of the hip may dislike this position and feel particularly unbalanced. You can get a sense of this yourself by standing with your feet together. Notice your adductors contracting to keep you in this position.

STEP 4   Muscle Bulk

Compare muscle bulk and the tone of the left and right thighs. Does the girth of the quadriceps appear equal?

What Your Findings Mean   As with other steps in which you observed muscle bulk, an increase in bulk suggests increased usage or weight bearing on that side, whereas atrophy of muscles (in a healthy person) suggests disuse. Atrophy in muscles of the lower limb is common following immobilisation of the limb or a prolonged period of bed rest.

STEP 5   Genu Valgum and Genu Varum

Next let's look at the knees. For this step, you need to ask your client to stand with the feet together, the medial malleoli as close together as possible. Is there evidence of genu valgum (a) or genu varum (b)?

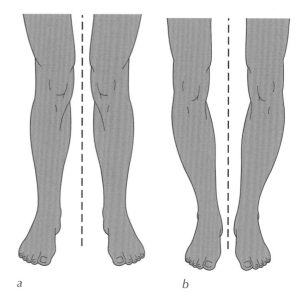

a   b

What Your Findings Mean   Genu valgum and genu varum affect both the knee joint itself and the muscles supporting it. Osteoarthritic changes or degradation of menisci may be more likely to occur on the side of the knee subject to greater compressive forces. Overstretching of soft tissues is likely on the opposite side of the knee.

In genu valgum, could increased pressure on the lateral side of the knee joint lead to degenerative changes on that side of the knee occurring before degenerative changes on the medial side of the knee? By contrast, in genu varum, there is increased pressure on the medial side of the knee joint. Structurally, with genu valgum, the muscles of the lateral thigh (the iliotibial band and biceps femoris) are shorter relative to the muscles of the medial side of the thigh (gracilis, semimembranosus and semitendinosus). Whereas with genu varum, the muscles of the medial side of the thigh (gracilis, semimembranosus and semitendinosus) are shorter relative to the muscles of the lateral thigh (iliotibial band and biceps femoris). Table 5.2 summarises this information.

Table 5.2   Changes to the Knee Joint and Surrounding Soft Tissues Corresponding With Genu Valgum and Genu Varum

	Genu valgum (knock knees)	Genu varum (bow legs)
Changes to knee joint	Increased pressure on the lateral side of the joint	Increased pressure on the medial side of the joint
Lengthened muscles	Gracilis Semimembranosus Semitendinosus	Iliotibial band Biceps femoris
Shortened muscles	Iliotibial band Biceps femoris	Gracilis Semimembranosus Semitendinosus

STEP 6   Patellar Position

The patella should be positioned in line with the tibial tuberosity. Look to see whether there is any maltracking of this bone. In these illustrations we show a right knee with lateral maltracking (a) and medial maltracking (b). Also, do the patellae seem to sit normally, or do they appear compressed and tilting against the knee joint?

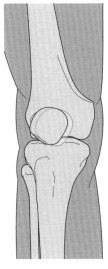 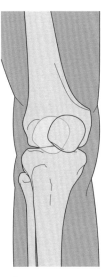

a   b

What Your Findings Mean   Because the patellae are housed within the quadriceps tendon, and this in turn is housed in fascia connected to other structures, could an increase in tension in the muscles or the fascia of the medial or lateral sides of the knee (or both) contribute to maltracking? For example, could lateral maltracking be due to increased tension in the lateral retinaculum of the knee and the iliotibial band? Could medial maltracking be due to increased tension in the vastus medialis?

Patellae that appear compressed against the knee joint are sometimes observed in clients who hyperextend their knees in standing, something you will have looked for in step 3 of your lower body lateral assessment (page 84). Anterior knee pain can sometimes be explained by patellae tilting such that their inferior poles stick into the fat pad beneath the knee, a condition perhaps aggravated by forced or prolonged knee extension.

STEP 7   Rotation at the Knee

The patella should point straight ahead with respect to the tibiofemoral joint. This means that if a client stands with the feet turned out slightly, as might be expected, the patella will also face outwards slightly, but should still be aligned over the joint. However, when there is rotation in the femur, the tibia or both, the patella no longer faces forwards.

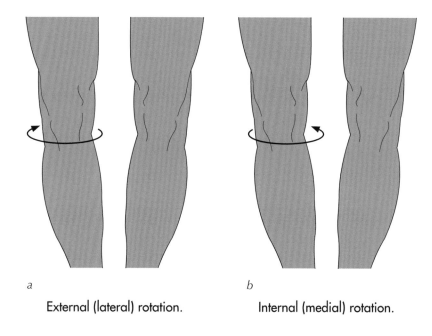

a

External (lateral) rotation.

b

Internal (medial) rotation.

TIP   One fun way to assess the knee anteriorly is to imagine that the patellae are the headlights of a car. Which way do the headlights shine? Where does their beam hit the ground?

What Your Findings Mean   A laterally rotated patella could correspond with a laterally rotated femur on that side, lateral tibial torsion or both. A medially rotated patella could correspond with a medially rotated femur on that side, medial tibial torsion or both. Clients who stand with the knees hyperextended often compress the patellae against the femurs, and the patellae slant downwards rather than facing straight ahead. Consequently, an imaginary headlight from the knees of these clients would illuminate the floor closer to the client than normal.

In addition to your postural assessment, you may want to carry out tests to confirm your diagnosis. A very simple test to assess for tibial torsion is to examine where the tibial tuberosity lies. This is in the midline of the anterior tibia, as you know, and will change direction with torsion of the bone.

Now that you are understanding the knees, take a look at these photographs and see if you can identify genu valgum of the left knee in figure *a*. Do you think this person is bearing weight equally through her left and right legs? In figure *b* the knees seem to squint inwards. (Note also the contour of the vastus medialis in this person.) The tibiae, feet and ankles of the client in *b* are all straight, facing forwards. Could this woman have internally rotated femurs? Look at the position of the person's right knee in figure *c*. Can you see that not only the knee but also the whole of the right lower limb appears externally rotated? Compare what sort of beam you think the knee 'headlights' would make in figures *a*, *b* and *c*.

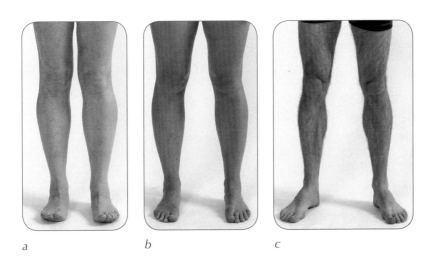

a   *b*   *c*

STEP 8   The Q Angle

Q angle describes the relationships among the pelvis, leg and foot. It measures the angle between the rectus femoris quadriceps muscle—hence the name Q angle—and the patellar tendon. It is useful because, theoretically, it may help you predict the likelihood of some types of knee problems and as such indicate the need for prophylactic treatment. To determine the Q angle of a client, follow these steps with the client standing:

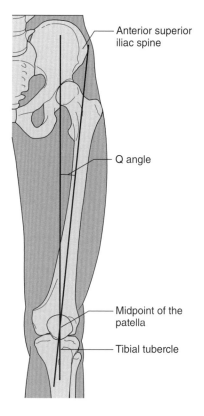

Anterior superior iliac spine

Q angle

Midpoint of the patella

Tibial tubercle

1. Find the midpoint of the patella.
2. From this point, draw a line running longitudinally up the femur to the ASIS (anterior superior iliac spine).
3. Find the tibial tuberosity.
4. Draw a line from the midpoint of the patella to the tibial tuberosity. Extend this line superior to the patella, thus creating an angle with your first line.

The angle between these two lines is the Q angle and is usually around 15 to 20 degrees, but it varies between males and females and among individuals.

It is more accurate to measure the Q angle of the client when standing than supine because when the client is standing, the patella is under the normal weight-bearing stresses.

What Your Findings Mean   Women have a greater Q angle than men do as a result of a wider pelvis. It has been postulated that when the Q angle is higher than normal, the client might experience greater stress through the patella when performing repetitive exercises that rely on the use of the knee. This could lead to maltracking of the patella so that it does not glide smoothly on the femoral grooves, which in turn could lead to microtrauma. Over time, this microtrauma could develop into a more serious pathology, such as degradation of the patellofemoral cartilage.

Clients with increased pronation of the foot may have an abnormal Q angle perhaps as a result of internal rotation of the tibia. If this rotation is prolonged, the alteration in normal biomechanics could again result in increased stress on the knee joint. This in turn could lead to more serious knee problems. It is important to remember, however, that an abnormal Q angle does not mean that a client *will* experience knee problems.

STEP 9 Tibia

Look at the leg now and compare the tibial tuberosities. Use them to determine whether there is any tibial torsion. There is usually a slight lateral rotation of the tibiae, corresponding with a turned-out foot position. Also look at the shape of the tibiae and whether either is bowed.

What Your Findings Mean   Bowing of the tibia could indicate osteomalacia or increased compressive forces on the concave side of the bone. Lateral tibial torsion produces the toe-out position and is associated with increased supination with the medial longitudinal arch of the foot being accentuated and the heel being inverted. Medial tibial torsion produces toe-in (pigeon-toed) feet, a decrease in the longitudinal arch, plus eversion of the heel. Table 5.3 summarises this information.

Table 5.3   Tibial Torsion and Corresponding Changes in the Foot

	Lateral tibial torsion	**Medial tibial torsion**
Overall foot position	Toe-out	Toe-in
Changes within the foot itself	There is increased supination, the heel is inverted, and the medial longitudinal arch is accentuated.	There is increased pronation, the heel is everted, and the medial longitudinal arch is decreased.

STEP 10   Ankles

When observing the ankles, the medial malleoli should be level with each other, and the lateral malleoli should be level with each other. Look also to see whether any swelling or discolouration is evident. Do you observe any eversion or inversion? In other words, does the client appear to be rolling in onto the medial side of the foot, or rolling out, with greater pressure on the outside of the foot and an increased space between the medial side of the foot and the floor?

What Your Findings Mean   Please refer to page 64, step 12 (malleoli) of the posterior postural assessment, for a full description of what changes in the position of malleoli might mean.

The ankles of the person shown here demonstrate how childhood musculoskeletal injuries can affect us for life. This 74-year-old woman fractured her left ankle very badly as a young girl.

STEP 11   Foot Position

How has your client positioned her feet? The feet should be positioned turned out to the same angle, equidistant from an imaginary plumb line.

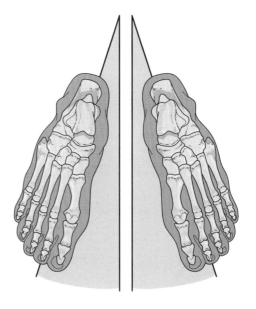

What Your Findings Mean   Feet turned out in a ballet-type stance could result from external rotation at the hip joint, lateral tibial torsion or both. External hip rotation could indicate shortening of the gluteus maximus and the posterior fibers of the gluteus medius, along with the iliotibial band. A client who stands with her feet turned inwards (pigeon-toed) may have shortened internal rotators of the hip, a medially rotated tibia or both. Table 5.4 summarises this information.

Table 5.4   Changes Associated With Toe-Out and Toe-In Foot Positions

	Toe-out position	Toe-in position
Possible position of the hip joint	Externally rotated	Internally rotated
Possible position of the tibia	Lateral tibial torsion	Medial tibial torsion
Muscles that might be shortened	External rotators of the femur; iliotibial band	Internal rotators of the femur

STEP 12   Pes Planus and Pes Cavus

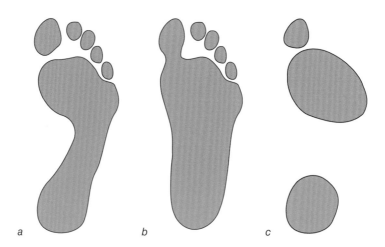

a   *b*   *c*

Weight should also appear to be distributed evenly between the medial and lateral aspects of each foot (*a*). Note whether there is pes planus (flat foot) (*b*) or pes cavus (hollow foot, or high arches) (*c*). In pes planus, the medial side of the foot might even be touching the floor completely, leaving no gap at all. In pes cavus, there will be a greater-than-normal space between the floor and the medial side of the foot.

TIP Many providers of sport footwear now have pressure plates on site to assess how potential buyers distribute their weight both in standing and running. A crude way to determine weight bearing in standing is simply to take footprints. Obviously, this is not something that is usually done as part of a postural assessment, but it is a fun activity to carry out at home to clarify your observations of family and friends.

What Your Findings Mean   Pes planus could result from weak intrinsic plantar muscles and an overextension of the corresponding ligaments of the foot leading to a fallen plantar arch. It corresponds with the pronation of the talus bone, which sometimes glides medially over the calcaneus. Over time the development of a flat foot might cause leg and foot pain as a result of overstretching of the long muscles of the sole of the foot. Pes cavus represents a higher longitudinal arch than normal. The calcaneus becomes supinated, and the remainder of the foot becomes pronated. Remember also that rotation of the trunk affects posture at the feet and ankles. Table 5.5 summarises this information.

Table 5.5   Changes Associated With Pes Planus and Pes Cavus

	Pes planus (flat foot)	Pes cavus (high arches)
Change in plantar arch	Loss of the plantar arch	Higher-than-normal plantar arch
Change in the position of the foot bones	The talus glides medially over the calcaneus.	The calcaneus supinates; the remainder of the foot pronates.
Change in soft tissues	Weakness in the intrinsic planar muscles; overstretching of the long muscles of the sole of the foot; overstretching of the ligaments and plantar fascia	Shortening of the intrinsic foot muscles and the plantar fascia
Relationship to trunk rotation	Trunk rotation to the left increases pronation on the right foot; trunk rotation to the right increases pronation in the left foot.	Trunk rotation to the left increases supination of the left foot and increased pressure through the lateral side of the left foot; trunk rotation to the right increases supination of the right foot and increased pressure through the lateral side of the right foot.

STEP 13   Other Observations

As in previous chapters in which you observed the body posteriorly and laterally, here you have an opportunity to note anything else that you have not yet documented. Pay particular attention to swelling around the joints, skin discolouration and scars. In this photograph, can you see the increase in tension in the tendon of the person's right tibialis anterior muscle? It could have been that she was correcting postural sway just as this photograph was taken, or she could have a significant difference between the tendons of this muscle on her left and right legs.

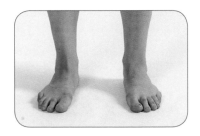

At the end of your assessment, stand back and take an overall view of your client. We all have unique body builds, known as somatotypes. There are three somatotypes: endomorph (a), ectomorph (b) and mesomorph (c). Commonly described as stocky or big boned, endomorphs have a large build, with greater fat deposits than the other two somatotypes. By contrast, ectomorphs are the slimmest of the three types, with prominent bony features and low body fat. They are commonly described as skinny or gangly. Mesomorphs are muscular, commonly described as athletic in appearance.

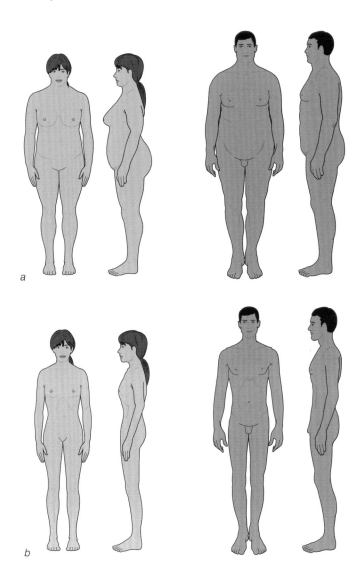

a

b

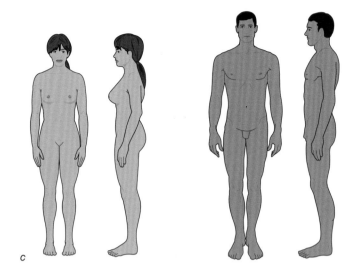

c

What Your Findings Mean   You may know of people who run marathons who are large and heavy, and you may know slim people who seem exceptionally strong. However, it is believed that certain body builds are more suitable to some physical activities than to others. This information can sometimes explain the increased likelihood for injury in some clients. For example, the ectomorphic-type body, with its slim physique and long limbs, does not lend itself to heavy weightlifting; the long levers of the limbs put joints and their ligaments at a disadvantage. Mesomorphs, by contrast, are larger and stockier, making their bodies more suited to weightlifting than perhaps to running. Although it is not good to overgeneralise, these observations may be useful at times—for example, when frequently injured clients appear to be participating in sports and activities not necessarily suited to their body types.

Quick Questions

1. When the clavicle is angled upwards quite sharply, what does this indicate?
2. What is the normal carrying angle of the elbow joint?
3. What are the common names for the knee positions genu valgum and genu varum?
4. Should there be slight lateral tibial torsion or slight medial tibial torsion in standing?
5. What do the terms *ectomorph, endomorph* and *mesomorph* mean in layperson's terms.

Seated Postural Assessment

This chapter focuses on the posture of clients who are clothed and seated. Although the assessment of seated posture is not normally carried out as part of the overall postural analysis, many people spend long hours sitting—at desks or driving, for example—so it is an important posture to understand. The information has been presented in this chapter in the usual step-by-step format, and many of the steps relate to those you may have already worked through in the assessment of the posterior and lateral postures. This chapter will help you assess clients who regularly maintain seated postures. If your client does not fall into this category, the information will reinforce your understanding of how the positioning of a joint affects the soft tissues that support it, when a person perpetuates a given position.

The information in this chapter is not intended to replace a full workstation analysis, something that should be carried out by a trained ergonomist. Nor is it intended to enable you to carry out an analysis of someone who regularly uses a wheelchair. If you are working with clients who use wheelchairs, the information will be helpful, but do bear in mind that it is based on the assessment of the general population and so some of the information will vary. Wheelchairs are provided for clients with wide-ranging physical postures, and these clients should be referred to physical therapists who specialise in working with this client group. This chapter concentrates on assessing the posture of a client seated at a desk, because many readers may be treating clients who have desk-based occupations.

You may read through this chapter for reference only, or you may locate the seated postural assessment chart on page 152 and work through it as you observe your client. Notice that I offer just two views, posterior and lateral. This is because you are unlikely to be observing the client anteriorly, either because he is seated at a desk and has something in front of him, or because he is seated in a vehicle.

Much of this information you will be familiar with if you have already worked through the chapters on posterior and lateral assessments. Note, too, that not all of the steps in the posterior and lateral assessments in earlier chapters are included. This is because here the client is clothed so you cannot observe bony landmarks, skin creases, and some of the joint positions. However, if you believe that your client's seated posture is contributing to the problem, you would be justified in carrying out a seated assessment with the client undressed and in a clinic environment, seated on a chair or stool.

It is obviously best to observe your client sitting at her workstation. The next best thing is to ask your client to assume the position she thinks she adopts at work, imagining that she is typing or using a computer mouse, for example. Ask her to demonstrate how she sits for the majority of the day, not just the position of good posture she takes up when she first starts work in the morning. The following photographs demonstrate the kinds of postures clients might adopt when they start work (a), when they are really concentrating on a screen in front of them (b) and at the end of the day (c). Notice what happens to the neck, even when the client is leaning back in the chair at the end of the day.

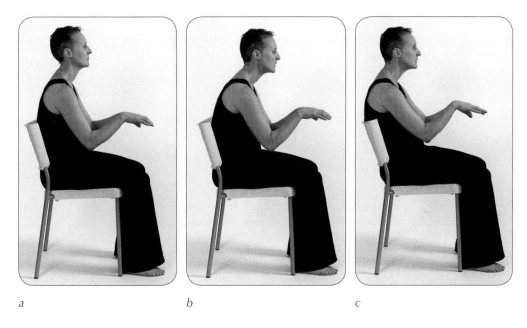

a   b   c

You may find that it is easier if you sit behind your client as you work through each of these steps.

STEP 1   Head and Neck Position

Begin by checking the position of the head and neck. Ask yourself whether the earlobes are level and whether there is there any lateral flexion in the neck. Is your client looking straight ahead, or can you see slightly more of one ear, or more of one side of the face than the other, indicating rotation to that side? Take a close look at the muscles of the cervical spine. Is there an increase in tone on either side? Does your client use a telephone a lot at work? If so, ask him to demonstrate: does he use a headset or wedge the telephone under one ear?

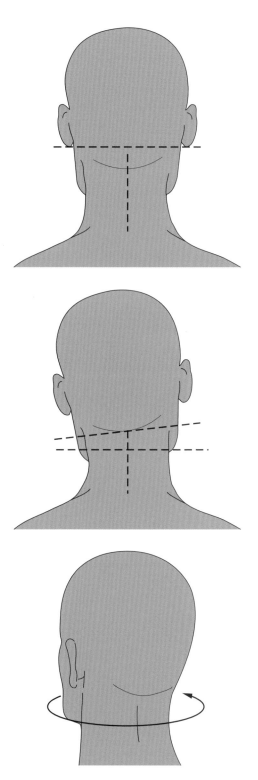

What Your Findings Mean   Uneven ears could indicate that the client has the head tilted to one side, flexing the cervical spine laterally to the side on which the ear is lower. Lateral neck flexion is also sometimes observed in clients with shoulder pain. To reduce pain, these clients often flex towards the painful side. Lateral flexion of the neck can result from shortened muscles on the side to which the neck is flexed. Rotation of the head might occur because a client positions some part of the workstation to one side. For example, he may have a keyboard in front of him but maintain his head rotated to the right to read from a document. Readers who are a bit rusty on their anatomy and physiology may find table 6.1 helpful.

Table 6.1   Muscles Shortened in Lateral Flexion or Rotation

	Lateral flexion	Rotation
To the right	Right levator scapulae Right sternocleidomastoid Upper fibers of the trapezius (on the right side)	Left sternocleidomastoid Right scalenes Right levator scapulae
To the left	Left levator scapulae Left sternocleidomastoid Upper fibers of the trapezius (on the left side)	Right sternocleidomastoid Left scalenes Left levator scapulae

STEP 2   Shoulder Height

Next, take a look at your client's shoulders. Are they level? How does your client position her arms? Does she rest them on the arms of her chair or on the desk, or does she rest them on a keyboard or on special arm supports? If driving, does she position both arms on the steering wheel, or does she drive using one hand predominantly? Does she rest one arm on the wheel and one arm on the windowsill? Although these questions concern the way a client uses her body and as such are not strictly postural, it is nevertheless important to know which positions clients perpetuate.

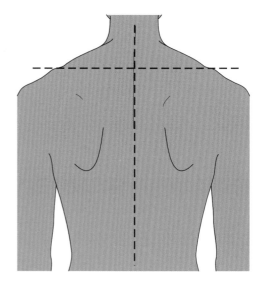

What Your Findings Mean   Because some shoulder muscles also attach to the neck, elevation of the shoulders is closely linked to head and neck posture. A client with a laterally flexed neck may also have an elevated shoulder on that same side. In step 8 on page 39 of the posterior postural assessment, you learned that if the scapula is elevated, you would expect the inferior angle of that scapula to be superior to the inferior angle of the scapula on the opposite side. However, because you cannot observe this in a clothed client, you may choose to palpate through the clothing for this structure instead.

Clients who rest one arm on the windowsill of a vehicle or on the arm of a chair are passively shortening the shoulder elevators on that side of the body. Could it be that because the muscles of the opposite arm are held isometrically, these clients sometimes experience problems in that shoulder rather than the one that is resting? Just as a client with shoulder pain might laterally flex the neck to that side, shortening the muscles on that side of the body in a protective manner, a client with neck pain might subconsciously elevate one or both shoulders to reduce the discomfort. This is another example of why it is necessary to take a case history before carrying out a postural assessment.

STEP 3   Thorax

Next, look at whether the client's workstation is directly in front of her or whether it is positioned to one side. Look at the client's chair and the position of her hips. Do the hips face forwards and the thorax another way?

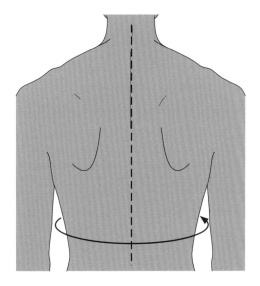

What Your Findings Mean   If a workstation is positioned just a little bit lateral to the direction in which the client's hips are facing, the client will need to rotate the thorax towards the workstation. Refer back to table 3.2 (page 44) to see more detail concerning the effects of pelvic rotation.

STEP 4   Hip and Thigh Position

Unless the client is seated on a stool, you may not be able to see the position of the hips and thighs when you are seated behind the client. Stand up and observe how he is sitting. Does he sit with his thighs close together, his feet and ankles neatly touching? Or, more likely, does he sit with his thighs abducted? Does he regularly sit with one leg crossed over the other?

What Your Findings Mean   Sitting for long periods of time with the hips abducted results in a lengthened and weakened gluteus maximus muscle but a shortened (and possibly also weakened) gluteus medius muscle. Many people frequently sit cross-legged to alleviate discomfort in the lumbar spine. If they always adopt the same position (e.g., placing the left leg over the right), this can lead to soft tissue changes and dysfunction.

STEP 5   Foot Position

The last step in this section is to look at the position of the client's feet. Does your client position her feet flat on the floor? Does she wear high heels, pushing her ankles into plantar flexion? Does she wrap her feet around the legs of a chair (as shown here), or sit cross-legged?

What Your Findings Mean   Wearing high heels results in shortening of the plantar flexors of the feet and ankles. Can you see from the photograph above how people who sit with their feet and ankles wrapped around the legs of a chair are in a position of eversion at the ankle and may therefore experience shortening of the fibular muscles? Try sitting in this position and notice that your hips internally rotate. This, too, may have consequences on posture.

Now, move your own position and sit so that you may observe your client's posture from the side.

STEP 1   Head and Neck Position

Ask yourself where the client's head is relative to the body. Does the head sit comfortably over the thorax, or is it pushed forwards? How does the cervical spine appear? Is the curve normal or flatter than normal, or is there increased lordosis? If you can see it, how does the cervicothoracic junction appear? Is C7 more prominent than usual?

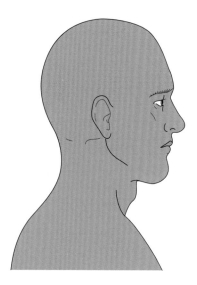

What Your Findings Mean   When the head is not positioned correctly over the thorax, the neck, chest and arms may all be affected. Maintenance of cervical flexion may result in increased tension in the scalenes and weakening and lengthening of the neck extensors. The increased strain placed on these muscles often results in pain not only in the neck but also in the shoulders and upper back.

Conversely, when the lordotic curve is exaggerated, cervical extensor muscles may become shortened and weak and neck flexors may become lengthened and weak. In this position there is increased compression on the posterior part of some of the cervical intervertebral discs. The zygapophyseal joints may be compressed, too. Have you come across clients who have perpetuated the forward head posture and report neurological symptoms? Could one explanation for these symptoms be nerve root compression as a result of this posture?

This forward head posture is also associated with an exaggerated kyphosis in the thoracic spine, so the thoracic cavity may be diminished.

STEP 2 Thorax

Clients who retain habitual static sitting postures often demonstrate an exaggeration in the normal thoracic curve. An increase in the thoracic region may be compensatory for an increase in cervical or lumbar lordosis or both.

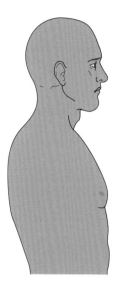

What Your Findings Mean   A severely kyphotic posture is associated with shortening of the anterior chest muscles and shallow breathing due to a depressed chest cavity. Sometimes there is shortening of the upper abdominals as a person slumps forwards, flexing the trunk. If the neck is lordotic, there may be weakness in the cervical spine flexors, the thoracic spine extensors, the middle and lower fibers of the trapezius and the rhomboids (due to protraction of the scapulae). The shoulder adductor muscles and the internal rotators may also be shortened when there is accompanying internal rotation of the humerus. Not surprisingly, neck and shoulder pain is common in clients with kyphotic postures because of these muscular imbalances.

STEP 3   Shoulder Position

Next, look at the shoulders. Although you will not get as much information from this observation as you did during the lateral postural assessment in standing, with the client unclothed, you can at least get a general feel here for the position of the shoulders relative to the head and neck. Do the shoulders appear protracted, the position associated with a kyphotic posture? Or, less likely, does the client sit erect with the chest out and the shoulders pulled back in military fashion? If driving a vehicle, are both shoulders flexed as the client holds the steering wheel?

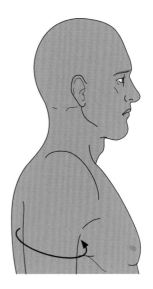

What Your Findings Mean   The slumped position many people adopt when sitting at a desk results in protraction of the shoulders and has consequences for the chest and neck as well as the glenohumeral joint itself. Because an increase in the curvature of the cervical spine often corresponds with this posture, the client may have shortened and weak cervical extensors.

In the chest region, protracted shoulders are associated with lengthened and weak rhomboids, tight pectoralis major and minor, and shortened intercostal muscles. The middle and lower fibers of the trapezius may be lengthened and weak also, as might the extensors of the thoracic spine. At the glenohumeral joint, an internally rotated humerus suggests shortness in muscles such as the subscapularis and teres major. In earlier chapters I asked whether this position might contribute to impingement syndromes resulting in pain on shoulder flexion.

STEP 4   Lumbar Spine, Pelvis and Hips

Some people sit with their legs abducted, or with one leg crossed over the other. In either case, in the sitting position the hips are flexed. Many people start their day with an upright posture, but as the muscles fatigue, posture changes and people become more kyphotic, with posteriorly tilted pelvises. Ask your clients to show you how they sit for most of the day.

What Your Findings Mean   When we sit upright, the lumbar spine is neutral and the pelvis is anteriorly tilted; when we slump, the lumbar spine flattens and the pelvis tilts posteriorly. When seated, many people cross one leg over the other. This is because if they try to maintain an upright posture, the pelvis tilts forwards, increasing lumbar lordosis. Crossing one leg over the other decreases pelvic tilt. In almost everyone who maintains a seated posture, the hip flexors are likely to become shortened.

You learned from chapter 4 (pages 78 and 81) that an increased lordotic curve indicates an anteriorly tilted pelvis and possibly results in lumbar pain perhaps from compression of the posterior part of the lumbar intervertebral discs and the zygapophyseal joints and shortening of the lumbar erector spinae. Yet most clients do not maintain this position throughout the day. The slumped position results in a decrease in the normal curve of the lumbar spine. In this case, soft tissues on the anterior of the body may be shortened, and the tissues of the posterior part of the lumbar spine may be stretched.

STEP 5   Knees

In the seated posture, the knees are always flexed unless the client is sitting on the floor with the legs straight out.

What Your Findings Mean   Flexed knees are associated with shortened knee flexors and lengthened knee extensors.

Quick Questions

1. Which muscles of the neck might be shortened or have increased tone in a client who has a workstation positioned to the right?

2. How do some people passively shorten the muscles that elevate the shoulder?

3. Anatomically speaking, what does crossing one leg over the other do to the lumbar spine and pelvis?

4. Which general group of hip muscles is always in a shortened position in the seated position?

5. What happens to the popliteus and the posterior of the knee joint in the seated position?

Postural Assessment Charts

You can use these postural assessment charts to track your findings and comments as you're completing postural assessments. Feel free to make photocopies of these charts for inclusion in your clients' records. The order of the steps in each of these charts matches the order of the steps in chapters 3, 4, 5 and 6. Consult those chapters for more detailed information on completing an assessment at each step.

POSTERIOR POSTURAL ASSESSMENT CHART

UPPER BODY

Left side		Right side
	Step 1 Ear Level	
	Step 2 Head and Neck Tilt	
	Step 3 Cervical Rotation	
	Step 4 Cervical Spine Alignment	
	Step 5 Shoulder Height	
	Step 6 Muscle Bulk and Tone	
	Step 7 Scapular Adduction and Abduction	
	Step 8 Inferior Angle of the Scapula	
	Step 9 Rotation of the Scapula	
	Step 10 Winging of the Scapula	

Left side		Right side
	Step 11 Thoracic Spine	
	Step 12 Thoracic Cage	
	Step 13 Skin Creases	
	Step 14 Upper Limb Position	
	Step 15 Elbow Position	
	Step 16 Hand Position	
	Step 17 Other Observations	
	LOWER BODY	
	Step 1 Lumbar Spine	
	Step 2 Pelvic Rim	
	Step 3 PSIS	

»continued

»continued

LOWER BODY

Left side		Right side
	Step 4 Pelvic Rotation	
	Step 5 Buttock Crease	
	Step 6 Thigh Bulk	
	Step 7 Genu Varum and Genu Valgum	
	Step 8 Posterior Knees	
	Step 9 Calf Bulk	

Left side		Right side
	Step 10 Calf Midline 	
	Step 11 Achilles Tendon Normal	
	Step 12 Malleoli 	
	Step 13 Foot Position 	
	Step 14 Other Observations	

LATERAL POSTURAL ASSESSMENT CHART

Right side			Left side
	Step 1 Head Position		
	Step 2 Cervical Spine		
	Step 3 Cervico- thoracic Junction		
	Step 4 Shoulder Position		
	Step 5 Thorax		
	Step 6 Abdomen		

UPPER BODY

Right side		Left side
	Step 7 Lumbar Spine	
	Step 8 Other Observations	
LOWER BODY		
	Step 1 Pelvis	
	Step 2 Muscle Bulk	
	Step 3 Knees	

»continued

»continued

LOWER BODY

Right side		Left side
	Step 4 Ankles 	
	Step 5 Feet 	
	Step 6 Other Observations	

From J. Johnson, 2012, *Postural assessment* (Champaign, IL: Human Kinetics).

ANTERIOR POSTURAL ASSESSMENT CHART

	UPPER BODY	
Right side		**Left side**

Right side		Left side
	Step 1 Face	
	Step 2 Head Position	
	Step 3 Muscle Tone	
	Step 4 Clavicles	
	Step 5 Shoulder Level	
	Step 6 Rounded Shoulders	
	Step 7 Chest	
	Step 8 Carrying Angle	

»continued

»continued

UPPER BODY

Right side		Left side
	Step 9 Arms	
	Step 10 Hands and Wrists	
	Step 11 Abdomen	

LOWER BODY

	Step 1 Lateral Pelvis	
	Step 2 Rotated Pelvis	
	Step 3 Stance	
	Step 4 Muscle Bulk	

Right side		Left side

	Step 5
	  Genu Valgum
	  and Genu Varum	

	Step 6 Patellar Position	

	Step 7
	  Rotation
	  at the Knee	

	Step 8 The Q angle	
	Step 9 Tibia	
	Step 10 Ankles	
	Step 11 Foot Position	

	Step 12
	  Pes Planus
	  and Pes Cavus	

	Step 13 Other Observations	

SEATED POSTURAL ASSESSMENT CHART

POSTERIOR VIEW

Left side		Right side
	Step 1 Head and Neck Position	
	Step 2 Shoulder Height	
	Step 3 Thorax	
	Step 4 Hip and Thigh Position	
	Step 5 Foot Position	

LATERAL VIEW

	Step 1 Head and Neck Position	
	Step 2 Thorax	

Left side		Right side

	Step 3
	  Shoulder
	  Position	

	Step 4 Lumbar Spine, Pelvis and Hips	
	Step 5 Knees	

Chapter 1

1. Factors that affect posture are structural or anatomical, age, physiological, pathological, occupation, hobbies and recreation, environmental, social and cultural, and mood and emotion.

2. Reasons for performing a postural assessment are to get more information, to save time, to establish a baseline, and to treat holistically.

3. A postural assessment might not be appropriate when treating an anxious client; a client unable to stand because of pain, illness, or instability; a client who does not understand the purpose of the assessment or who does not give consent to having one performed; or a client who would benefit from a different form of assessment, more appropriate to his condition (e.g., Parkinson's disease or following a stroke).

4. In most cases, it is important to take a medical history before carrying out a postural assessment because information may be revealed that affects whether the assessment is appropriate and safe.

5. When analysing various parts of the body and how they fit together, it important to always take an overall view of the client because all parts are interrelated. Patients dislike being referred to as 'the knee' or 'the shoulder'.

Chapter 2

1. Useful equipment to have when carrying out a postural assessment includes a warm, private room; a full-length mirror; body crayons (and cleansing wipes); postural assessment charts; and a model skeleton.

2. Bony landmarks that are useful to identify before starting a posterior postural assessment include the medial border of the scapula, the inferior angle of the scapula, the spinous processes of the spine, the olecranon process of the elbow, the posterior superior iliac spine (PSIS), knee creases, the midline of the calf, and the midline of the Achilles tendon.

3. Any of the questions included in table 2.1 are suitable starting points for postural assessment.

4. A neutral pelvis is one in which the left and right iliac crests, left and right PSIS, and left and right ischia are level when the client is viewed posteriorly, and in which the ASIS and pubis are in the same vertical plane when the client is viewed laterally.

5. Possible contraindications to postural assessment include an inability to stand or sit because of pain, blood pressure issues, and poor balance. Check for allergies if using body crayons to mark bony landmarks.

Chapter 3

1. The right sternocleidomastoid, levator scapulae, scalenes and upper fibers of the trapezius all laterally flex the neck to the right.

2. Atrophy of shoulder muscles may result from immobility of the upper limb and conditions such as adhesive capsulitis (frozen shoulder).

3. *Winging* is a term often used to describe the way the inferior angle (and often the medial border) of the scapula tilts away from the rib cage, becoming prominent. True winging involves damage to serratus anterior or the long thoracic nerve.

4. Lateral flexion to the left and a left elevated pelvis suggests a shortened left quadratus lumborum muscle.

5. The midline of the calf might appear more lateral on one leg if the hip of that side is internally rotated or if the tibia on that side is rotated inwards with respect to the femur—or if both conditions exist.

Chapter 4

1. A forward head posture might increase the strain placed on the muscles of the posterior neck such as the levator scapulae, resulting in pain in the neck, shoulders and upper back.

2. Muscles that become shortened when the humerus is internally rotated include the subscapularis, teres major and pectoralis major.

3. Retaining static postures such as sitting at a desk or driving for long periods contribute to an increased kyphosis in the thorax.

4. When the pelvis tilts anteriorly, there is an increase in the lordotic curve of the lumbar spine.

5. A client who stands with flexed knees is likely to have shortened hamstrings.

Chapter 5

1. A steep incline in the angle of the clavicle indicates elevated shoulders and tension in the muscles associated with shoulder elevation.

2. The normal carrying angle of the elbow is 5 to 10 degrees in males and 10 to 15 degrees in females.

3. The common name for genu valgum is knock kneed; the common name for genu varum is bow legged.

4. There should be slight lateral tibial torsion in standing.

5. Endomorphs are commonly described as stocky or big boned; ectomorphs are described as skinny or gangly; and mesomorphs are described as athletic or muscular.

Chapter 6

1. When a client has a workstation positioned to the right, the muscles of the neck that might be shortened or have increased tone are the left sternocleidomastoid, right levator scapulae and right scalenes.

2. Some people passively shorten the muscles that elevate the shoulder by resting that arm on the windowsill of a vehicle or on the arm of a chair.

3. Crossing one leg over the other overcomes the anterior tilting of the pelvis and the increase in lumbar lordosis associated with this posture.

4. Hip flexor muscles are always shortened in the seated position.

5. Assuming that a client sits on a regular chair (and not with the legs outstretched, knees in extension), the soft tissues of the posterior knee, including the popliteus muscle, maintain a shortened position.

References

Anderson J.E., ed. 1978. *Grant's Atlas of Anatomy*. Baltimore/London: Williams & Wilkins.

Cloward, R.B. 1959. "Cervical Diskography." *Annals of Surgery* 150: 1052-1064.

Earls, J., and T. Myers. 2010. *Fascial Release for Structural Balance*. Chichester, UK: Lotus, and Berkeley, CA: North Atlantic Books.

Green, Walter B., and James D. Heckman, eds. 1993. *The Clinical Measurement of Joint Motion*. Rosemont, IL: American Academy of Orthopaedic Surgeons.

Hanchard, N., L. Goodchild, J. Thompson, T. O'Brien, C. Richardson, D. Davison, H. Watson, M. Wragg, S. Mtopo, and M. Scott. 2011. "Evidence-Based Clinical Guidelines for the Diagnosis, Assessment and Physiotherapy Management of Contracted (Frozen) Shoulder v.1.3, 'Standard' Physiotherapy." Endorsed by the Chartered Society of Physiotherapy.

Hertling, D., and R.M. Kessler. 1996. *Management of Common Musculoskeletal Disorders*. Philadelphia: Lippincott.

Johnson, G., N. Bogduk, A. Nowitzke, and D. House. 1994. "Anatomy and Actions of the Trapezius Muscle." *Clinical Biomechanics* 9: 44-50.

Kendall, F. P., E.K. McCreary, and P.G. Provance. 1993. *Muscles Testing and Function*. Baltimore: Williams & Wilkins.

Kendall, H.O., F.P. Kendall, and D.A. Boynton. 1952. *Posture and Pain*. Baltimore: Williams & Wilkins.

Levangie, P.K., and C.C. Norkin. 2001. *Joint Structure and Function: A Comprehensive Analysis*. Philadelphia: Davis.

Magee, David J. 2002 *Orthopaedic Physical Assessment*. Philadelphia: Saunders.

Myers, T. 2001. "Psoas." *Massage and Bodywork*. February/March, April/May, June/July and August/September.

Schleip, R. 2008. *The Nature of Fascia* (DVD).

Jane Johnson, MSc, is co-director of the London Massage Company, England. As a chartered physiotherapist and sport massage therapist, she has been carrying out postural assessments for many years.

Johnson teaches postural assessment as a provider of continuing professional development (CPD) workshops for the Federation of Holistic Therapists (FHT). This experience has brought her into contact with thousands of therapists of all disciplines and helped inform her own practice. She is also a regular presenter at the annual Complementary and Massage Expo (CAM) held in the United Kingdom.

Johnson is a full member of the Chartered Society of Physiotherapists and is registered with the Health Professions Council. A member of the Institute of Anatomical Sciences, she has a deep interest in musculoskeletal anatomy and how newly qualified therapists can be better educated in this subject. She also is interested in the relationship between emotions and posture.

In her spare time, Johnson enjoys taking her dog for long walks, practicing wing chun kung fu, and visiting museums. She resides in London.

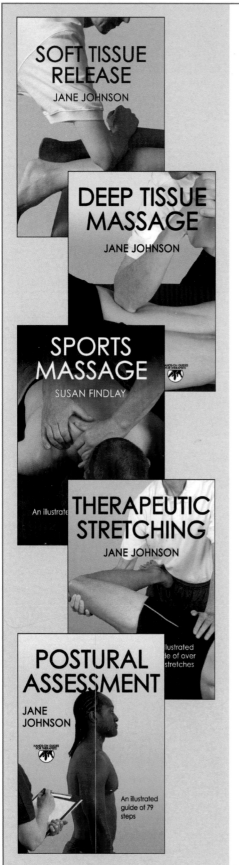